职业教育云计算类课程系列教材

Linux集群与运维技术

郑美容　朱晓彦　顾旭峰◎主　编
黄金凤　胡　晶　程伍端◎副主编

中国铁道出版社有限公司
CHINA RAILWAY PUBLISHING HOUSE CO., LTD.

内容简介

本书参照 Linux 相关国际认证教材编写方式，参考 "HX" 职业技能等级认证云计算平台运维开发相关内容，采用知识内容和典型任务案例相结合的结构，通过 9 章内容对主流的 Linux 系统架构与运维技术进行了全面的介绍，所有内容均经过实际验证，并配有实训资源包和源代码。

本书可作为 "1+X" 云计算平台运维与开发职业资格认证相关内容的融通教材，也可作为高职高专、职业本科、应用型本科的计算机网络技术、云计算技术应用、大数据技术应用等计算机类相关专业的教材及相关 Linux 中、高级运维人员的技术参考书。

图书在版编目（CIP）数据

Linux 集群与运维技术 / 郑美容，朱晓彦，顾旭峰主编. —北京：
中国铁道出版社有限公司，2022.6
职业教育云计算类课程系列教材
ISBN 978-7-113-28937-9

Ⅰ.①L… Ⅱ.①郑… ②朱… ③顾… Ⅲ.①Linux操作系统-职业教育-教材 Ⅳ.①TP316.85

中国版本图书馆CIP数据核字（2022）第037178号

书　　名：Linux 集群与运维技术
作　　者：郑美容　朱晓彦　顾旭峰

策　　划：翟玉峰　　　　　　　　　　　　编辑部电话：（010）83517321
责任编辑：翟玉峰　许　璐
封面设计：郑春鹏
责任校对：孙　玫
责任印制：樊启鹏

出版发行：中国铁道出版社有限公司（100054，北京市西城区右安门西街8号）
网　　址：http://www.tdpress.com/51eds/
印　　刷：三河市兴博印务有限公司
版　　次：2022年6月第1版　2022年6月第1次印刷
开　　本：787 mm×1 092 mm　1/16　印张：16.5　字数：421千
书　　号：ISBN 978-7-113-28937-9
定　　价：45.00元

版权所有　侵权必究

凡购买铁道版图书，如有印制质量问题，请与本社教材图书营销部联系调换。电话：（010）63550836
打击盗版举报电话：（010）63549461

前 言

从20世纪90年代初至今，Linux操作系统为IT行业做出了巨大的贡献。随着虚拟化、云计算、大数据时代的到来，Linux更是得以飞速发展，占据了整个服务器行业的半壁江山，但同时它也面临着巨大的挑战。当今互联网企业多样化的需求、高难度复杂的业务需要及不断扩展的应用需求，都需要越来越合理的管理模式来保证Linux服务器的安全、稳定、高可用性，而这些维护离不开Linux运维工作人员的付出。

本书主要根据行业趋势，由浅入深地介绍常用的Linux系统架构和运维技术。全书分为9章：第1章Linux基础运维与Shell脚本，介绍基本的运维命令和脚本的编写语法；第2章LAMP与LNMP，介绍Web站点的常见架构和部署调试方法；第3章数据库运维与集群，介绍基础数据库管理运维及主流读写分离的主从数据库架构与实现；第4章NoSQL数据库服务，介绍NoSQL缓存数据库的部署与运维；第5章共享存储与分布式存储，介绍NFS、GlusterFS、Ceph和Swift四种常见的存储架构和运维；第6章集群与高可用技术，介绍常见的Keepalived、LVS和HAProxy集群和负载均衡软件的部署与运维；第7章监控服务与自动化运维工具，介绍Nagios和Zabbix监控工具及Ansbile自动化运维工具；第8章容器高级运维，介绍容器的相关部署与运维；第9章Kubernetes服务架构与平台，介绍主流的Kubernetes容器平台的应用与运维。

本书参照Linux相关国际认证教材编写方式，每章由知识内容和典型的实训案例相结合，所有内容均经过实际验证，并配有实训资源包和源代码。

本书参考"1+X"云计算平台运维与开发职业资格认证相关内容，通过任务案例的方式进行撰写，难度适中、内容丰富，对主流的Linux系统架构与运维技术进行了全面的介绍，可以作为"1+X"云计算平台运维与开发职业资格认证相关内容的融通教材，也可以作为高等职业院校计算机网络技术、云计算技术应用、大数据技术应用等计算机类相关专业的教材及相关Linux中高级运维人员的技术参考书。

本书由具有丰富教学和实践经验的院校老师与行业专家联合编写，三位主编共同设计教材内容架构。主编郑美容、朱晓彦是"1+X"云计算平台运维与开发职业资格认证金牌讲师，具有多年Linux教学经验，顾旭峰是南京第五十五所技术开发有限公司"1+X"云计算平台运维与开发职业技能等级标准及认证体系专家，具有丰富的企业技术经验，相关编写教师均为专业一线教师。本书具体分工如下：第1、2、3章由朱晓彦编写；第4章由黄金凤编写；第5章由胡晶编写；第6章由程伍端编写；第7、8、9章由郑美容编写。江苏一道云科技发展有限公司、南京第五十五所技术开发有限公司参与了案例的设计、提供工作，蔡晨亮、李双霖参与了本书的开发与验证工作，在此一并表示感谢。

由于编者水平有限，疏漏和不足之处在所难免，恳请读者批评、指正。

编　者

2021 年 9 月

目 录

第1章 Linux基础运维与Shell脚本 1

1.1 Linux概述 2
　1.1.1 Linux操作系统简介 2
　1.1.2 如何学习Linux操作系统 3
1.2 Linux运维命令 3
　1.2.1 执行帮助文档命令 3
　1.2.2 常用系统工作命令 6
　1.2.3 系统状态检测命令 8
　1.2.4 文本文件编辑命令 11
　1.2.5 文件目录管理命令 12
1.3 Shell脚本语言 13
　1.3.1 vim文本编辑器 14
　1.3.2 Shell基础知识 15
　1.3.3 Shell常用功能 16
　1.3.4 任务：编写简单的Shell脚本 18
1.4 开放研究任务：Shell脚本项目实施 19
小结 27

第2章 LAMP与LNMP 28

2.1 LAMP服务 29
　2.1.1 LAMP服务的原理 29
　2.1.2 LAMP服务的架构 29
2.2 LNMP服务 31
　2.2.1 LNMP服务的原理 31
　2.2.2 LNMP服务的架构 32
2.3 LAMP服务的部署 32
　2.3.1 LAMP服务解析 32

　2.3.2 任务：部署LAMP服务 33
2.4 开放研究任务：用LNMP架构部署WordPress网站 40
小结 49

第3章 数据库运维与集群 50

3.1 数据库服务概述 51
　3.1.1 数据库服务背景 51
　3.1.2 数据库服务原理 52
3.2 数据库服务运维 52
　3.2.1 初始化MariaDB服务 53
　3.2.2 MariaDB数据库运维 54
3.3 主从复制数据库 60
　3.3.1 主从复制数据库服务 60
　3.3.2 任务：部署主从复制数据库 61
3.4 读写分离数据库 63
　3.4.1 读写分离数据库服务 63
　3.4.2 Mycat中间件服务 64
　3.4.3 任务：配置Mycat数据库中间件 65
3.5 开放研究任务：构建读写分离数据库集群 67
小结 75

第4章 NoSQL数据库服务 76

4.1 NoSQL数据库服务概述 77
　4.1.1 NoSQL服务背景 77
　4.1.2 NoSQL服务简介 77
4.2 Memcached服务 78

4.2.1 Memcached服务简介 79	第6章 集群与高可用技术 126
4.2.2 Memcached服务运维 80	6.1 Keepalived服务 127
4.2.3 任务：部署Memcached	6.1.1 Linux集群 127
服务 82	6.1.2 Keepalived服务及原理 127
4.3 Redis服务 84	6.1.3 任务：部署Keepalived
4.3.1 Redis服务简介 84	高可用集群 130
4.3.2 Redis服务运维 86	6.2 LVS服务 135
4.3.3 任务：部署Redis服务 87	6.2.1 负载均衡 135
4.4 开放研究任务：用Sentinel实现	6.2.2 LVS服务及原理 136
Redis集群高可用部署 91	6.2.3 任务：部署DR模式的LVS
小结 95	负载均衡 138
	6.3 HAProxy服务 142
第5章 共享存储与分布式存储 96	6.3.1 负载均衡类型 142
5.1 NFS共享存储 97	6.3.2 HAProxy服务 143
5.1.1 NFS共享存储概述 97	6.3.3 任务：部署HAProxy负载
5.1.2 NFS工作原理 97	均衡集群 144
5.1.3 任务：部署NFS共享存储 98	6.4 开放研究任务：搭建Linux
5.2 GlusterFS分布式存储 101	集群架构之负载均衡 148
5.2.1 GlusterFS分布式存储概述 101	小结 155
5.2.2 GlusterFS工作原理 101	
5.2.3 任务：部署GlusterFS	第7章 监控服务与自动化运维工具 156
分布式存储 102	7.1 Nagios监控服务 157
5.3 Ceph分布式存储 110	7.1.1 Nagios服务 157
5.3.1 Ceph分布式存储概述 110	7.1.2 Nagios工作原理 157
5.3.2 Ceph工作原理 111	7.1.3 Nagios监控部署 158
5.3.3 任务：安装Ceph服务 113	7.2 Zabbix监控服务 164
5.4 Swift分布式存储 115	7.2.1 Zabbix服务和原理 164
5.4.1 Swift分布式存储概述 115	7.2.2 Zabbix服务安装与运维 167
5.4.2 Swift服务架构 116	7.2.3 任务：部署Zabbix分布式
5.4.3 任务：使用Swift作为	监控系统 171
后端存储 118	7.3 Ansible自动化运维 175
5.5 开放研究任务：架构Ceph	7.3.1 Ansible服务 175
分布式存储集群 120	7.3.2 Ansible任务执行 176
小结 125	

7.3.3 任务：Ansible部署Nginx
服务 177
7.4 开放研究任务：Ansible批量
部署LAMP环境 181
小结 185

第8章 容器高级运维 186

8.1 容器技术 187
    8.1.1 Docker容器技术 187
    8.1.2 Docker服务原理 188
    8.1.3 Docker服务架构 189
    8.1.4 任务：Docker引擎的安装 191
8.2 Docker容器管理 194
    8.2.1 Docker仓库概述 194
    8.2.2 Docker镜像的管理和使用 195
    8.2.3 Docker容器的管理和使用 197
    8.2.4 Docker存储与网络的使用 198
    8.2.5 任务：Docker容器仓库
配置 201
8.3 Dockerfile编写 203
    8.3.1 Dockerfile简介 203
    8.3.2 Dockerfile指令 204
    8.3.3 任务：Dockerfile编写案例 206
8.4 Docker容器编排 210
    8.4.1 Dockcr容器编排简介 210

8.4.2 容器编排管理 211
8.4.3 任务：docker-compose
编排 212
8.5 开放研究任务：容器应用
商城系统编排 215
小结 224

第9章 Kubernetes服务架构与平台 ... 225

9.1 认识Kubernetes服务 226
    9.1.1 Kubernetes背景 226
    9.1.2 Kubernetes原理 226
    9.1.3 Kubernetes架构 228
9.2 Kubernetes集群服务 230
    9.2.1 Kubernetes集群概念 230
    9.2.2 Kubernetes集群管理 231
    9.2.3 任务：Kubernetes集群
部署 233
9.3 Kubernetes编排服务 240
    9.3.1 编排文件的概念和语法 240
    9.3.2 资源编排与资源文件编写 243
    9.3.3 任务：Kubernetes编排
案例 247
9.4 开放研究任务：Kubernetes
编排WordPress 250
小结 256

第 1 章

Linux 基础运维与 Shell 脚本

本章概要

学习 Linux 操作系统中的基础运维命令,并掌握 Shell 编程,完成运维脚本的编写。

学习目标

- 学习 Linux 操作系统,了解 Linux 历史与发展。
- 通过学习掌握 Linux 基础运维命令。
- 通过学习掌握 Shell 编程的方法。
- 使用 Linux 操作系统与 Shell 编程脚本完成实验。

思维导图

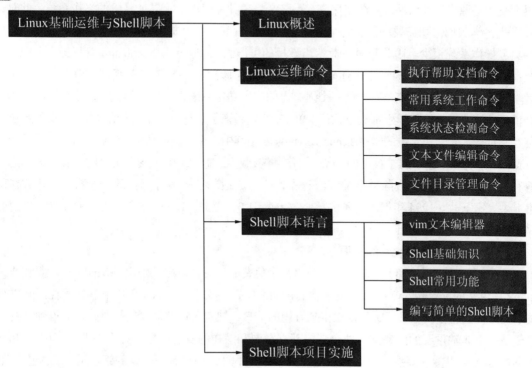

任务目标

在 Linux 操作系统中使用基础命令和 Shell 命令构建运维脚本。

1.1 Linux 概述

学习目标

学完本节后，您应能够：
- 了解什么是 Linux。
- 了解 Linux 的产生与发展。
- 了解如何学习 Linux 操作系统。

1.1.1 Linux 操作系统简介

1. Linux 操作系统介绍

Linux 是一套免费使用和自由传播的类 UNIX 操作系统，是一个基于 POSIX 和 UNIX 的多用户、多任务、支持多线程和多 CPU 的操作系统。Linux 能运行主要的 UNIX 工具软件、应用程序和网络协议，支持 32 位和 64 位硬件。Linux 继承了 UNIX 以网络为核心的设计思想，是一个性能稳定的多用户网络操作系统。

2. Linux 的产生与发展

Linux 操作系统的诞生、发展和成长过程始终依赖着五个重要支柱：UNIX 操作系统、MINIX 操作系统、GNU 计划、POSIX 标准和 Internet 网络。

20 世纪 80 年代，计算机硬件的性能不断提高，PC 的市场不断扩大，当时可供计算机选用的操作系统主要有 UNIX、DOS 和 MacOS 这几种。UNIX 价格昂贵，不能运行于 PC；DOS 显得简陋，且源代码被软件厂商严格保密；MacOS 是一种专门用于苹果计算机的操作系统。此时，计算机科学领域迫切需要一个更加完善、强大、廉价和完全开放的操作系统。由于供教学使用的典型操作系统很少，因此当时在荷兰当教授的美国人 Andrew S.Tanenbaum 编写了一个操作系统，名为 MINIX，用于向学生讲述操作系统内部工作原理。MINIX 虽然很好，但只是一个用于教学目的的简单操作系统，而不是一个强有力的实用操作系统，其最大的优点就是公开源代码。全世界学计算机专业的学生都通过钻研 MINIX 源代码来了解计算机中运行的 MINIX 操作系统，芬兰赫尔辛基大学二年级的学生 Linus Torvalds 就是其中一个。在吸收了 MINIX 精华的基础上，Linus 于 1991 年写出了属于自己的 Linux 操作系统，版本为 Linux 0.01，是 Linux 时代开始的标志。他利用 UNIX 的核心，去除繁杂的核心程序，改写成适用于一般计算机的 x86 系统，并放在网络上供大家下载。1994 年推出完整的核心 Version 1.0，至此，Linux 逐渐成为功能完善、稳定的操作系统，并被广泛使用。

3. Linux 与 Windows 的不同

在中国，Windows 和 Linux 对个人用户都是免费的，对企业用户来说，Windows 需要收费，Linux 是免费的。Windows 提供了较完善的帮助系统，而大多数 Linux 用户依靠技术社区来进行技术支持，他们可以从聊天版和论坛得到有用的信息，缺点是往往需要较长时间才能得到答案。安全是一个非常重要的问题。这两个系统都有漏洞和潜在的安全问题。许多管理员声称，Linux 比 Windows 有更多的漏洞，这当然跟它开源有关。不过，如果有一个好的管理员，安全应该不是一个重大问题。无论使用哪种操作系统，开源就是指对外部开放软件源代码。Linux 是开源的，而 Windows 并不开源。

1.1.2 如何学习 Linux 操作系统

首先，当前学习 Linux 操作系统是不错的选择，一方面 Linux 操作系统的应用范围比较广泛，尤其在大数据、物联网和人工智能领域有大量的应用场景；另一方面 Linux 操作系统是开源的，可以通过阅读其核心源代码来深入了解操作系统的体系结构和资源管理方式。在当前的大数据、人工智能时代背景下，不仅 IT（互联网）行业的职场人应该学习 Linux 操作系统，传统行业的职场人学习 Linux 操作系统也有一定的积极意义。

学习 Linux 操作系统通常要经历三个阶段，每个学习阶段有不同的学习方式和学习目标：

第一个阶段是熟悉 Linux 操作系统的应用，掌握 Linux 操作系统的安装和运行方式。这个阶段的学习难度相对比较低，也完全可以通过自学来掌握 Linux 操作系统的基础应用。

第二个阶段是基于 Linux 操作系统来完成一些行业应用，如云计算就是一个不错的选择。其中，通过 Linux 来完成一些虚拟化操作，也会为理解云计算奠定一个扎实的基础。而在这个过程中，有助于系统了解 Linux 操作系统的资源管理方式。Linux 在云计算领域的应用是非常普遍的，而且很多学习 Linux 操作系统的初学者未来也可以从事云计算运维类岗位。

第三个阶段是基于 Linux 操作系统进行一些程序设计。这个阶段要了解 Linux 操作系统的资源管理方式和程序调度方式，要学习一系列 Linux 的 API，这个过程可以看成是深入学习 Linux 操作系统的重要阶段。

1.2　Linux 运维命令

学习目标

学完本节后，您应能够：
- 掌握执行帮助文档命令。
- 掌握常用系统工作命令。
- 掌握系统状态检测命令。
- 掌握文本文件编辑命令。
- 掌握文件目录管理命令。

1.2.1　执行帮助文档命令

1. 内建命令与外部命令

内建命令实际上是 Shell 程序的一部分，其中包含的是一些比较简单的 Linux 系统命令，这些命令是写在 Bash 源码的 builtins 里面的，由 Shell 程序识别，并在 Shell 程序内部完成运行。通常在 Linux 系统加载运行时，Shell 就被加载并驻留在系统内存中。而且解析内部命令 Shell 不需要创建子进程，因此其执行速度比外部命令快，如 history、cd、exit 等。

外部命令是 Linux 系统中的实用程序部分，因为实用程序的功能通常都比较强大，所以其包含的程序量也会很大，在系统加载时并不随系统一起被加载到内存中，而是在需要时才将其调用内存。通常外部命令的实体并不包含在 Shell 中，但是其命令执行过程是由 Shell 程序控制的。Shell 程序管理外部命令执行的路径查找、加载存放，并控制命令的执行。外部命令是在 Bash 之外额外安装的，通常放在 /bin、/usr/bin、/sbin、/usr/sbin 等中。可通过 echo $PATH 命令查看外部命令的存储路径，如 cd、ls 等。

```
[root@localhost ~]# type cd
cd is a shell builtin
[root@localhost ~]# type ls
ls is aliased to 'ls --color=auto'
[root@localhost ~]# type mkdir
mkdir is /usr/bin/mkdir
```

得到这样的结果说明是内建命令，正如上文所说，内建命令都是在 Bash 源代码中的 builtins 的 .def 中。

cd is a shell builtin 意思是得到这样的结果说明是外部命令，正如上文所说，外部命令在 /usr/bin 或 /usr/sbin 等中。

2. 执行查看帮助命令

（1）help 作为内部命令的用法。

```
[root@localhost ~]# help cd
cd: cd [-L|[-P [-e]]] [dir]
    Change the shell working directory.

    Change the current directory to DIR.  The default DIR is the value of the
    HOME shell variable.

    The variable CDPATH defines the search path for the directory containing
    DIR.  Alternative directory names in CDPATH are separated by a colon (:).
    A null directory name is the same as the current directory.  If DIR begins
    with a slash (/), then CDPATH is not used.

    If the directory is not found, and the shell option 'cdable_vars' is set,
    the word is assumed to be  a variable name.  If that variable has a value,
    its value is used for DIR.

    Options:
        -L      force symbolic links to be followed
        -P      use the physical directory structure without following symbolic
    links
        -e      if the -P option is supplied, and the current working directory
    cannot be determined successfully, exit with a non-zero status

    The default is to follow symbolic links, as if '-L' were specified.

    Exit Status:
    Returns 0 if the directory is changed, and if $PWD is set successfully when
    -P is used; non-zero otherwise.
```

（2）help 作为外部命令的用法。

```
[root@localhost ~]# cd --help
-bash: cd: --: invalid option
cd: usage: cd [-L|[-P [-e]]] [dir]
```

(3) man 命令的用法。

man 命令可以通过一些参数，快速查询 Linux 帮助手册，并且格式化显示。并且 man 命令没有内建与外部命令的区分。

语法：

```
man [-adfhktwW] [section] [-M path] [-P pager] [-S list] [-m system] [-p string] title..
```

man 命令常用参数用法，见表 1-1，man 命令其他参数用法，见表 1-2。

表 1-1 man 命令常用参数

参　　数	参　数　解　释
-a	显示所有匹配项
-d	显示 man 查找手册文件的时候，搜索路径信息，不显示手册页内容
-D	同 -d，显示手册页内容
-f	同命令 whatis，将在 whatis 数据库查找以关键字开头的帮助索引信息
-h	显示帮助信息
-k	同命令 apropos，将搜索 whatis 数据库，模糊查找关键字
-S list	指定搜索的领域及顺序，如：-S 1:1p httpd
-t	使用 troff 命令格式化输出手册页，默认：groff 输出格式页
-w	不带搜索 title 打印 manpath 变量。带 title 关键字，打印找到手册的文件路径，默认搜索一个文件后停止
-W	同 -w
section	搜索领域"限定手册类型"，默认查找所有手册

表 1-2 man 命令其他参数

参　　数	参　数　解　释
-c	显示使用 cat 命令的手册信息
-C	指定 man 命令搜索配置文件，默认：man.config
-K	搜索一个字符串在所有手册页中，速度很慢
-M	指定搜索手册的路径
-P pro	使用程序 pro 显示手册页面，默认：less
-B pro	使用 pro 程序显示 HTML 手册页，默认：less
-H pro	使用 pro 程序读取 HTML 手册，用 txt 格式显示，默认：cat
-p str	指定通过 groff 格式化手册之前，先通过其他程序格式化手册

示例如下所示：

```
[root@localhost ~]# man ls
    -a, --all
          do not ignore entries starting with .
    -A, --almost-all
          do not list implied . and ..
    --author
          with -l, print the author of each file
    -b, --escape
```

```
              print C-style escapes for nongraphic characters
       --block-size=SIZE
              scale sizes by SIZE before printing them; e.g., '--block-
size=M' prints sizes in units of 1,048,576 bytes; see SIZE format below
       -B, --ignore-backups
              do not list implied entries ending with ~
       -c     with -lt: sort by, and show, ctime (time of last modification
of file status information); with -l: show ctime and sort by name; otherwise:
              sort by ctime, newest first
       -C     list entries by columns
       --color[=WHEN]
              colorize the output; WHEN can be 'never', 'auto', or 'always'
 (the default); more info below
       -d, --directory
              list directories themselves, not their contents
```

打开手册之后可以通过鼠标滚轮或上下键来进行上下翻看,使用【PgUp】和【PgDn】键进行上下翻页。使用"/关键字"命令来查找关键字,此时按【n】键,会向下匹配关键字,按【N】键会向上匹配关键字。

(4) info 命令的用法。

info 格式的帮助命令,比 man 以及 help, info 更为实用。语法如下所示:

```
info(选项)(参数)
```

选项:
- -d:添加包含 info 格式帮助文档的目录。
- -f:指定要读取的 info 格式的帮助文档。
- -n:指定首先访问的 info 帮助文件的节点。
- -o:输出被选择的节点内容到指定文件。

参数:
- 帮助主题:指定需要获得帮助的主题,可以是命令、函数以及配置文件。

常用快捷键:
- 【?】键,显示 info 的常用快捷键。
- 【N】键:显示(相对于本节点的)下一节点的文档内容。
- 【P】键:显示(相对于本节点的)前一节点的文档内容。
- 【U】键:进入当前命令所在的主题。
- 【M】键:按【M】键后输入命令的名称,就可以查看该命令的帮助文档了。
- 【G】键:按【G】键后输入主题名称,进入该主题。
- 【L】键:回到上一个访问的页面。
- 【Space】键:向前滚动一页。
- 【Backup】或【Del】键:向后滚动一页。
- 【Q】键:退出 info。

1.2.2 常用系统工作命令

1. ls 命令

ls 命令用于列出目录以及所含文件信息及子目录。

所处的工作目录不同，当前工作目录下的文件肯定也不同。使用 ls 命令的"-a"参数可以看到全部文件（包含隐藏文件），使用"-l"参数可以查看文件的属性、大小等详细信息。将这两个参数结合之后，再执行 ls 命令即可查看当前目录中的所有文件的属性信息：

```
[root@localhost ~]# ls  -al
total 40
dr-xr-x---.  3 root root 4096 May 24 08:24 .
dr-xr-xr-x. 17 root root 4096 Apr 18  2017 ..
-rw-------.  1 root root  550 May 24 07:44 .bash_history
-rw-r--r--.  1 root root   18 Dec 29 02:26 .bash_logout
-rw-r--r--.  1 root root  176 Dec 29 02:26 .bash_profile
-rw-r--r--.  1 root root  176 Dec 29 02:26 .bashrc
-rw-r--r--.  1 root root  100 Dec 29 02:26 .cshrc
-rw-------   1 root root   35 May 24 08:24 .lesshst
```

如果想查看目录属性信息，则需要额外添加一个 -d 参数。例如，可以使用如下命令查看 /opt 目录的属性与权限信息：

```
[root@localhost ~]# ls -ld /opt
drwxr-xr-x. 2 root root 6 Jun 10  2021 /opt
```

2. cd 命令

cd 命令用于切换当前工作目录至目标目录，命令后面可添加相对路径与绝对路径。

这个命令属于最常用的一个 Linux 命令。可以通过 cd 命令迅速地切换到不同的工作目录。除了常见的切换目录功能之外，还可以使用"cd ~"命令切换到当前用户的家目录，以及使用"cd -"往后退一个工作路径。例如，进入 /opt 目录，然后退回来：

```
[root@localhost ~]# cd /opt/
[root@localhost opt]# cd -
/root
```

3. echo 命令

echo 命令用于在终端输出字符串或者变量提取后的值。例如，把指定的字符串"Hello Linux！"输出到终端屏幕的命令为：

```
[root@localhost ~]# echo Hello Linux!
Hello Linux!
```

4. pwd 命令

pwd 命令用于显示用户当前所处的目录。如果用户不知道当前所处的目录，就必须使用它来查看。例如，查看当前用户所处目录：

```
[root@localhost ~]# pwd
/root
```

5. date 命令

date 命令用于显示及设置系统的时间或日期。

只需要在强大的 date 命令中输入以"+"号开头的参数，即可按照指令格式来输出系统的时间或日期，这样在日常工作中便可以把备份数据的命令与指令格式输出的时间信息结合到一起。例如，按照默认格式查看当前系统时间的 date 命令如下所示：

```
[root@localhost ~]# date
Sat May 24 22:34:28 UTC 2021
```

按照 "年-月-日 时:分:秒" 格式查看当前系统的时间 date 命令如下所示：

```
[root@localhost ~]# date "+%y-%m-%d %H:%M:%S"
21-02-24 22:36:33
```

6. find 命令

find 命令用来在指定目录下查找文件。任何位于参数之前的字符串都将被视为欲查找的目录名。如果使用该命令时，不设置任何参数，则 find 命令将在当前目录下查找子目录与文件。并且将查找到的子目录和文件全部进行显示。

```
[root@localhost ~]# find
.
./.bash_logout
./.bash_profile
./.bashrc
./.cshrc
./.tcshrc
./.ssh
./.ssh/authorized_keys
./.ssh/id_rsa.pub
./.ssh/id_rsa
./.bash_history
./.viminfo
./.lesshst
```

使用 find 命令查看权限为 777 的文件。

```
[root@localhost ~]# find / -perm 777
/usr/bin/rvi
/usr/bin/rview
/usr/bin/view
/usr/bin/lastb
/usr/bin/zsoelim
/usr/bin/iptables-xml
```

1.2.3 系统状态检测命令

作为一名合格的运维人员，想要更快、更好地了解 Linux 服务器，就需要具备快速查看 Linux 系统运行状态的能力，因此接下来会讲解关于 Linux 服务器的网络、内核、负载、内存等命令。

1. ip 命令

ip 命令和 ifconfig 类似，但前者功能更强大，并旨在取代后者。使用 ip 命令，只需一个命令，就能很轻松地执行一些网络管理任务。ifconfig 是 net-tools 中已被废弃使用的一个命令，许多年前就已经没有维护了。iproute2 套件里提供了许多增强功能的命令，ip 命令即是其中之一。

例如，使用 ip 命令查看当前系统的网络信息，命令如下所示：

```
[root@localhost ~]# ip a
1: lo: <LOOPBACK,UP,LOWER_UP> mtu 65536 qdisc noqueue state UNKNOWN
    link/loopback 00:00:00:00:00:00 brd 00:00:00:00:00:00
    inet 127.0.0.1/8 scope host lo
       valid_lft forever preferred_lft forever
    inet6 ::1/128 scope host
       valid_lft forever preferred_lft forever
2: eth0: <BROADCAST,MULTICAST,UP,LOWER_UP> mtu 1450 qdisc pfifo_fast state UP qlen 1000
```

```
link/ether fa:16:3e:df:85:63 brd ff:ff:ff:ff:ff:ff
inet 10.0.0.5/24 brd 10.0.0.255 scope global dynamic eth0
   valid_lft 76075sec preferred_lft 76075sec
inet6 fe80::f816:3eff:fedf:8563/64 scope link
   valid_lft forever preferred_lft forever
```

2. ifconfig 命令

ifconfig 命令用于获取网卡配置与网络状态等信息，使用 ifconfig 命令来查看本机当前的网卡配置与网络状态等信息时，其实主要查看的就是网卡名称、inet 参数后面的 IP 地址、ether 参数后面的网卡物理地址等。

```
[root@localhost ~]# ifconfig
eth0: flags=4163<UP,BROADCAST,RUNNING,MULTICAST>  mtu 1450
        inet 10.0.0.5  netmask 255.255.255.0  broadcast 10.0.0.255
        inet6 fe80::f816:3eff:fedf:8563  prefixlen 64  scopeid 0x20<link>
        ether fa:16:3e:df:85:63  txqueuelen 1000  (Ethernet)
        RX packets 3749  bytes 331418 (323.6 KiB)
        RX errors 0  dropped 0  overruns 0  frame 0
        TX packets 3329  bytes 691758 (675.5 KiB)
        TX errors 0  dropped 0 overruns 0  carrier 0  collisions 0

lo: flags=73<UP,LOOPBACK,RUNNING>  mtu 65536
        inet 127.0.0.1  netmask 255.0.0.0
        inet6 ::1  prefixlen 128  scopeid 0x10<host>
        loop  txqueuelen 0  (Local Loopback)
        RX packets 6  bytes 416 (416.0 B)
        RX errors 0  dropped 0  overruns 0  frame 0
        TX packets 6  bytes 416 (416.0 B)
        TX errors 0  dropped 0 overruns 0  carrier 0  collisions 0
```

3. uname 命令

uname（uname 的全称为 unix name）命令用于显示系统信息。uname 可显示计算机以及操作系统的相关信息。

```
[root@localhost ~]# uname
Linux
[root@localhost ~]# uname -a
Linux localhost 3.10.0-229.el7.x86_64 #1 SMP Fri Mar 6 11:36:42 UTC 2015 x86_64 x86_64 x86_64 GNU/Linux
```

4. free 命令

free 命令用于显示内存状态。free 指令会显示内存的使用情况，包括实体内存，虚拟的交换文件内存，共享内存区段，以及系统核心使用的缓冲区等。

```
[root@localhost ~]# free -h
```

执行 free -h 命令后的输出信息见表 1-3。

表 1-3　执行 free -h 命令后的输出信息

类型	内存总量 （total）	已用量 （used）	可用量 （free）	进程共享的内存量 （shared）	磁盘缓存的内存量 （buff/cache）	缓存的内存量 （available）
Mem:	3.9 GB	94 MB	3.6 GB	16 MB	166 MB	3.6 GB
Swap:	8.0 GB	0 B	8.0 GB			

5. df 命令

df 命令用于显示目前在 Linux 系统上的文件系统的磁盘使用情况。

```
[root@localhost ~]# df -h
Filesystem      Size  Used Avail Use% Mounted on
/dev/vda1        60G  911M   60G   2% /
devtmpfs        2.0G     0  2.0G   0% /dev
tmpfs           2.0G     0  2.0G   0% /dev/shm
tmpfs           2.0G   17M  2.0G   1% /run
tmpfs           2.0G     0  2.0G   0% /sys/fs/cgroup
```

6. who 命令

who 命令用于显示目前系统中有哪些使用者，显示的资料包含使用者 ID、使用的终端机、从哪边连上来的、上线时间、呆滞时间、CPU 使用量、动作等。

```
[root@localhost ~]# who
root     pts/0        2021-02-24 22:34 (172.16.180.225)
```

7. history 命令

history 命令用于显示历史记录和执行过的指令命令。history 命令读取历史命令文件中的目录到历史命令缓冲区和将历史命令缓冲区中的目录写入命令文件。该命令单独使用时，仅显示历史命令，在命令行中，可以使用符号！执行指定序号的历史命令。例如，要执行第 2 个历史命令，则输入 !2。

在内存中，历史命令仅能存储 1 000 条历史命令，该数量由环境变量 HISTSIZE 进行控制。默认不显示命令的执行时间，history 已经记录该时间，只是没有显示。

```
[root@localhost ~]# history
1  pwd
2  clear
3  date
4  date "+%y-%m-%d %H:%M:%S"
5  ip a
6  ifconfig
7  clear
8  find
9  find / -perm 777
10  df
11  time ls
12  uname
13  uname -a
14  free
15  free -h
16  who
17  last
18  history
```

注意：

如想查询某个用户在系统上执行了什么命令，可以使用 root 用户身份登录系统，检查 Home 目录下用户主目录下的 ".bash_history" 文件，该文件记录了用户所使用的命令和历史信息。

1.2.4 文本文件编辑命令

1. cat 命令

cat 命令主要用于滚屏显示文件内容或将多个文件合并成一个文件。通常使用 cat 命令查看文件内容，但是 cat 命令的输出内容不能分页显示，要查看超过一屏的文件内容，需要使用 more 或 less 等其他命令。如果在 cat 命令中没有指定参数，则 cat 会从标准输出中获取内容。例如，查看 file1 文件内容的命令为：

```
[root@localhost ~]# echo 这是第一个文件 > file1
[root@localhost ~]# cat file1
这是第一个文件
```

利用 cat 命令还可以合并多个文件。例如，要把 file1 和 file2 文件的内容合并为 file3，且文件的内容在 file1 的前面，命令为：

```
[root@localhost ~]# echo 这是第二个文件 > file2
[root@localhost ~]# cat file2 file1 > file3
[root@localhost ~]# cat file3
这是第二个文件
这是第一个文件
```

2. more 命令

more 命令类似 cat，不过会以一页一页的形式显示，更方便使用者逐页阅读，而最基本的指令就是按【Space】键，即显示下一页；按【b】键（back）就会往回一页显示；按【h】键可查看使用中的说明文件；按【q】键退出 more 命令；而且还有搜寻字串的功能（与 vi 相似）。例如，以分页方式查看 file1 的文件内容。

```
[root@localhost ~]# more file1
这是第一个文件
[root@localhost ~]# cat file1 | more
这是第一个文件
```

3. less 命令

less 与 more 类似，但使用 less 可以随意浏览文件，而 more 仅能向前移动，不能向后移动，而且 less 在查看之前不会加载整个文件。例如，以分页方式查看 file1 的文件内容。

```
[root@localhost ~]# less file1
这是第一个文件
file1 (END)
```

4. head 命令

head 命令用于显示文本文档的前 N 行，默认情况下只显示文件的前 10 行内容。

```
[root@localhost ~]# history | head
    1  clear
    2  help cd
    3  cd --help
    4  help ls
    5  ls --help
    6  help cd
    7  cd
    8  cd /opt
    9  cd -
   10  echo Hello Linux!
```

5. tail 命令

tail 命令可以将文件指定位置到结束位置的内容写到标准输出。使用 tail 命令的 -f 选项可以方便地查阅正在改变的日志文件。tail -f filename 会把文件里最尾部的内容显示在屏幕上，并且不断刷新，以便用户查看最新的文件内容。

```
[root@localhost ~]# tail -f /var/log/messages
May 25 07:01:01 controller systemd: Starting Session 38 of user root.
May 25 08:01:01 controller systemd: Started Session 39 of user root.
May 25 08:01:01 controller systemd: Starting Session 39 of user root.
May 25 08:19:18 controller systemd-logind: Removed session 35.
May 25 08:24:16 controller systemd-logind: Removed session 37.
```

6. wc 命令

wc 命令用于统计指定文件中的字节数、字数、行数，并将统计结果显示输出。该命令统计指定文件中的字节数、字数、行数。如果没有给出文件名，则从标准输入读取。wc 同时也给出所指定文件的总统计数。

```
[root@localhost ~]# wc file1
 1  1 22 file1
[root@localhost ~]# wc file1 file2          // 文件批量查询信息
 1  1 22 file1
 1  1 22 file2
 2  2 44 total                              // 此行为两个文件的总用量
```

7. diff 命令

diff 命令用于比较文件的差异。diff 命令分析两个文件，并输出两个文件不同的行。diff 的输出结果表明需要对一个文件做怎样的操作之后才能与第二个文件相匹配。diff 并不会改变文件的内容，但是 diff 可以输出一个 ed 脚本来应用这些改变。现在看一下 diff 是如何工作的，假设有两个文件：

```
[root@localhost ~]# diff file1 file2
1c1
< 这是第一个文件
---
> 这是第二个文件
```

使用带有 -c 参数的 diff 命令来描述文件内容具体的不同之处：

```
[root@localhost ~]# diff -c file1  file3
*** file1           2021-02-24 23:54:23.839440399 +0000
--- file3           2021-02-24 23:56:25.188798452 +0000
***************
*** 1 ****
--- 1,2 ----
+ 这是第二个文件
  这是第一个文件
```

1.2.5 文件目录管理命令

在 Linux 操作系统的日常运维工作中，还需要掌握对文件的创建、修改、复制、剪切、更名与删除等操作。

1. touch 命令

touch 命令用于修改文件或者目录的时间属性，包括存取时间和更改时间。若文件不存在，系统会建立一个新的文件。

在创建空白的文本文件方面，这个 touch 命令相当简捷，例如，touch Linux 命令可以创建一个名为 Linux 的空白文本文件。

```
[root@localhost ~]# touch Linux
[root@localhost ~]# ls -l
total 12
-rw-r--r-- 1 root root   0 May 25 02:54 Linux
```

2. mkdir 命令

mkdir 命令用于创建空白的目录，格式为"mkdir [选项] 目录"。除了能创建单个空白目录外，mkdir 命令还可以结合 -p 参数来递归创建文件目录。

```
[root@localhost ~]# mkdir /opt/file
[root@localhost ~]# mkdir -p /opt/file/file2
[root@localhost ~]# cd /opt/file
[root@localhost file]# cd file2/
[root@localhost file2]#
```

3. cp 命令

cp 命令主要用于文件或目录的复制，格式为"cp [参数] 源文件 目标文件"。

```
[root@localhost ~]# touch install.log
[root@localhost ~]# cp install.log test.log
[root@localhost ~]# ls
install.log  test.log
```

4. mv 命令

mv 命令用于剪切文件或将文件重命名，格式为"mv [选项] 源文件 [目标路径 | 目标文件名]"。剪切操作不同于复制命令，因为它会默认把源文件删除，只保留剪切后的文件。

```
[root@localhost ~]# mv test.log file1.log
[root@localhost ~]# ls
file1.log  install.log
```

5. rm 命令

rm 命令用于删除目录或文件，格式为"rm [参数] 文件"。在 Linux 操作系统中删除文件时，系统会默认再次确认是否要执行删除操作，如果不想看到此提问，可以添加 -f 参数进行系统的强制性删除。

```
[root@localhost ~]# rm file1.log
rm: remove regular empty file 'file1.log'? y
[root@localhost ~]# rm -f install.log
[root@localhost ~]# ls
```

1.3 Shell 脚本语言

学习目标

学完本节后，您应能够：
- 掌握 vim 文本编辑器。
- 了解 Shell 基础知识。
- 掌握 Shell 常用功能。
- 编写 Shell 简单脚本。

1.3.1 vim 文本编辑器

1. vim 文本编辑简介

vim 是由 vi 发展演变过来的文本编辑器，因其具有语法高亮显示、多视窗编辑、代码折叠、支持插件等功能，现已成为众多 Linux 发行版本的标配。对于初学者来说，vim 往往是生涩、难以学习的文本编辑器，但当你完全掌握了这种编辑器之后，你会发现自己的工作效率往往会比没有使用 vim 之前提升了许多倍。

2. vim 工作模式

vim 具有多种工作模式，常用的工作模式有：普通模式、插入模式、命令模式。普通模式主要用于光标的移动，文字迅速定位等光标相关大量的快捷键操作。插入模式顾名思义，可以实现文本的基本编辑功能。命令模式用于文件的处理，如文件的保存，退出，显示行号等。

直接输入 vim 命令，即可开启该文本编辑器，默认将创建一个新的文档（因为没有指定文件名，所以保存时需要提供文件名）。另外，如果 vim 命令后跟了文件名参数，则需要判断该文件是否存在，如果存在，vim 将打开该文件，如果不存在，vim 将创建该文件。

3. vim 命令参数

（1）vim 插入模式。

插入模式可通过以下参数进入，见表 1-4。

表 1-4　vim 插入模式命令参数

按　键	功　能　描　述
a	进入插入模式，后续输入的内容将插入至当前光标的后面
A	进入插入模式，后续输入的内容将插入至当前段落的段尾
i	进入插入模式，后续输入的内容将插入至当前光标的前面
I	进入插入模式，后续输入的内容将插入至当前段落的段首
o	进入插入模式并在当前行的后面创建新的空白行
O	进入插入模式并在当前行的前面创建新的空白行

当需要退回普通模式或不知道自己当前处于什么模式时，可以通过【Esc】键返回普通模式。

（2）vim 常用命令。

在 vim 编辑器中编辑文档内容主要有两种常用方式：进入编辑模式操作和快捷键操作。快捷键操作方式是在普通模式下输入相应的快捷键实现对应的功能。快捷键功能描述参见表 1-5。

表 1-5　vim 常用命令

命　令	功　能　描　述
x	删除光标当前字符
dd	删除（剪切）光标所在整行
5dd	删除（剪切）从光标处开始的 5 行
yy	复制光标所在整行
n	显示搜索命令定位到下一个字符串
N	显示搜索命令定位到上一个字符串
u	撤销上一步操作
p	粘贴至当前行之后

（3）vim 保存与退出。

一般情况下，我们会通过命令模式输入特定的指令实现保存与退出功能，常用指令见表 1-6。

表 1-6　vim 保存与退出

命　　令	功　能　描　述
:w	保存
:q	退出
:q!	强制退出
:wq!	强制保存退出
:set nu	显示行号
:命令	执行该命令
/字符串	在文本中从上至下

1.3.2　Shell 基础知识

1. Shell 简介

Shell 本身是一个用 C 语言编写的程序，它是用户使用 UNIX/Linux 的桥梁，用户的大部分工作都是通过 Shell 完成的。Shell 既是一种命令语言，又是一种程序设计语言。作为命令语言，它交互式地解释和执行用户输入的命令；作为程序设计语言，它定义了各种变量和参数，并提供了许多在高级语言中才具有的控制结构，包括循环和分支。它虽然不是 UNIX/Linux 系统内核的一部分，但它调用了系统核心的大部分功能来执行程序、建立文件并以并行的方式协调各个程序的运行。因此，对于用户来说，Shell 是最重要的实用程序，深入了解和熟练掌握 Shell 的特性及其使用方法，是用好 UNIX/Linux 系统的关键。可以说，Shell 使用的熟练程度反映了用户对 UNIX/Linux 使用的熟练程度。

Shell 有两种执行命令的方式：①交互式（Interactive），即解释执行用户的命令，用户输入一条命令，Shell 就解释执行一条；②批处理（Batch），即用户事先写一个 Shell 脚本（Script），其中有很多条命令，让 Shell 一次把这些命令执行完，而不必一条一条地输入命令。

Shell 脚本和编程语言很相似，也有变量和流程控制语句，但 Shell 脚本是解释执行的，不需要编译，Shell 程序从脚本中一行一行读取并执行这些命令，相当于一个用户把脚本中的命令一行一行输入到 Shell 提示符下执行。Shell 初学者请注意，在平常应用中，建议不要用 root 账号运行 Shell。作为普通用户，不管是有意还是无意，都无法破坏系统；但如果是 root，那就不同了，只要敲几个字母，就可能导致灾难性后果。

2. Shell 的分类

程序设计语言大致可分为两类：编译型语言和解释型语言。

（1）编译型语言：很多传统的程序设计语言（如 Fortran、Ada、Pascal、C、C++ 和 Java）都是编译型语言。这类语言需要预先将写好的源代码（source code）转换成目标代码（object code），这个过程称为"编译"。运行程序时，直接读取目标代码。由于编译后的目标代码非常接近计算机底层，因此执行效率很高，这是编译型语言的优点。但是，由于编译型语言多半运作于底层，所处理的是字节、整数、浮点数或其他机器层级的对象，往往实现一个简单的功能需要大量复杂的代码。

（2）解释型语言：解释型语言也被称为"脚本语言"。执行这类程序时，解释器（interpreter）需要读取我们编写的源代码，并将其转换成目标代码，再由计算机运行。因为每次执行程序都多了编译的过程，因此效率有所下降。使用脚本编程语言的好处是，它们多半运行在比编译型语言还高的层级，能够轻易处理文件与目录之类的对象；缺点是它们的效率通常不如编译型语言。不过权衡之下，通常使用脚本编程还是值得的：花一个小时写成的简单脚本，同样的功能用 C 或 C++ 来编写实现，可能需要两天，而且一般来说，脚本执行的速度已经够快了，快到足以让人忽略它性能上的问题。脚本编程语言的例子有 AWK、Perl、Python、Ruby 与 Shell。

1.3.3 Shell 常用功能

1. Shell 脚本格式

每一个完善的脚本都要遵循一些既定的规则，Shell 脚本也是一样，有自己独特的格式，下面来看一个完善的 Shell 脚本，分析其格式风格。脚本文件 example.sh 的内容如下所示：

```
#!/bin/bash                              //指定脚本程序的命令解释器
# Date:2021-2-15                         //指定编写脚本的时间
# Version:0.1                            //指定版本号
# This program is system test            //解释脚本的作用
# Author:baicai                          //指定脚本的作者
echo "Hello Linux!"                      //Shell 脚本程序内容
```

2. 流程控制语句

Shell 有一套自己的流程控制语句，其中包括条件语句（if），循环语句（for,while），选择语句（case）。本节通过实例介绍各个语句的使用方法。

现有一段 Shell 脚本代码如下所示：

```
c_netinfo(){
CARDNUM=0
MODENUM=0
CARDNAME=ens33
BOOTPT="dhcp"
while[[ $CARDNUM != 1 ]] && [[ $CARDNUM != 2 ]]
do
CARDNUM=1
echo -n "Plz enter the network device:[1(ens33),2(ens34),default 1]:"
read CARDNUM
if[[ $CARDNUM==1 ]];
then
        CARDNAME=ens33
          break
elif[[ $CARDNUM==2 ]];
then
        CARDNAME=ens34
        CARDNUM=1
          break
elif[[ $CARDNUM=="" ]];
then
        CARDNAME=ens33
          break
else
        echo "Plz input choice 1 or 2[1(ens33),2(ens34),default 1]:"
fi
done
```

（1）while 条件循环语句。

while 条件循环语句是一种让脚本根据某些条件来重复执行命令的语句，while 循环语句通过判断条件测试的真假来决定是否继续执行命令。

```
while [[ $CARDNUM != 1 ]] && [[ $CARDNUM != 2 ]]
// 如果变量 CARDNUM 的值不等于 1 且不等于 2 就执行循环
do
```

```
CARDNUM=1
// 设定该变量默认值为 1
echo -n "Plz enter the network device:[1(ens33),2(ens34),default 1]:"
// 输出提示让用户选择输入 1 或 2, 默认为 1
read CARDNUM
// 识别刚刚输入的数字, 并调给变量 CARDNUM
done
循环结束
```

(2) if 条件测试语句。

if 条件测试语句可以让脚本根据实际情况自动执行相应的命令。

```
if [[ $CARDNUM == 1 ]];            // 如果 CARDNUM 变量为 1, 则
then
        CARDNAME=ens33             // 把 ens33 赋值给 CARDNAME 变量
        break                      // 退出循环
elif [[ $CARDNUM == 2 ]];          // 否则如果 CARDNUM 变量为 2, 则
then
        CARDNAME=ens34             // 把 ens34 赋值给 CARDNAME 变量
        CARDNUM=1                  // 还原 CARDNUM 变量默认值为 1
        break                      // 退出循环
elif [[ $CARDNUM == "" ]];         // 否则如果 CARDNUM 为空, 则
then
        CARDNAME=ens33             // 把 ens33 赋值给 CARDNAME 变量
        break                      // 退出循环
else
        echo "Plz input choice 1 or 2[1(ens33),2(ens34),default 1]:"
// 否则输出提示让用户选择输入 1 或 2, 默认为 1
fi
```

(3) case 条件测试语句。

Shell case 语句为多选择语句。可以用 case 语句匹配一个值与一个模式, 如果匹配成功, 执行相匹配的命令。

在以下 Shell 脚本中外层使用了 for 语句进行无限循环, 内部使用 case 语句进行匹配。

```
for ((i=1;;i++))                   // 外层使用 for 语句进行无限循环
do
menu
case "$choice" in                  // 如果变量 choice 值为 1, 则执行 c_netinfo 函数
"1")
c_netinfo
;;
"2")                               // 如果变量 choice 值为 2, 则执行 c_hostname 函数
c_hostname
;;
"q")                               // 如果变量 choice 为 q, 则退出循环体
exit 0
;;
esac
done
```

3. 正则表达式

正则表达式, 又称正规表示式、正规表示法、针对表达式、规格表达式、常规表达法 (Regular

Expression，或 regex、regexp、RE）。正则表达式使用单个字符来描述、匹配一系列符合某个句法规则的字符串。在很多编辑器里，正则表达式通常用来检索、替换那些符合某个模式的文本。

正则表达式中有些匹配字符与 Shell 中的通配符号一样，但含义却不同。如下是部分代码：

```
MY_IP1='ip a show dev ens33 | grep -w inet | awk '{print $2}''
```

使用 ip 命令查看 ens33 网络，并使用 grep -w 匹配 inet 那一行，使用 awk 命令提取第二个字段，最后赋值给 MY_IP1。

```
GW='route -n | grep UG | awk '{print $2}''
```

使用 route -n 命令查看路由表，并使用 grep 筛选出 UG 的行，使用 awk 命令提取第二个字段，最后赋值给 GW。

```
echo "ens33:   IP:$MY_IP1"
```

输出 ens33 的 IP 为上一步提取的变量信息。

1.3.4 任务：编写简单的 Shell 脚本

在 Shell 脚本中不仅会用到前面学习过的很多 Linux 命令以及正则表达式、管道符等语法规则，还需要把内部功能模块化通过逻辑语句进行处理，最终形成日常所见的 Shell 脚本。

任务执行清单

在本任务中，您将通过所学 Shell 知识，编写一个更改主机名和主机配置文件的 Shell 脚本。

目标

- 学会 Shell 使用方法。
- 掌握 Shell 脚本编写。

重要信息

- 本任务采用 CentOS 7 操作系统。
- 虚拟机密码为 000000。

解决方案

1. 编写 Shell 脚本

进入安装好的 Linux 操作系统，使用 vim 命令进行 Shell 脚本编写。代码如下所示：

```
[root@controller ~]# vim host.sh
#!/bin/bash
# Date:2021-2-15
# Version:0.1
# Author:baicai
LANG=C;export LANG

echo -n "Plz input the hostname:"
read hostname
hostnamectl set-hostname $hostname
sed -i '3,$d' /etc/hosts
echo "$IPADDR $hostname" >> /etc/hosts
#while [[ $HOSTNUM != n ]] || [[ $HOSTNUM == "" ]]
while [[ True ]]
do
```

```
echo -n "Do you want to add hosts:[y or n,default no]:"
read HOSTNUM
[[ $HOSTNUM == "" ]] && HOSTNUM="n"
case "$HOSTNUM" in
"y")
    echo -n "plz enter hostname in hosts:"
    read hostname
    echo -n "plz enter hostname ip in hosts:"
    read IPADDR1
    echo "$IPADDR1 $hostname" >> /etc/hosts
;;
"n")
break
;;
*)
;;
esac
done
cat /etc/hosts
```

2. 运行 Shell 脚本

```
[root@controller ~]# bash host.sh
Plz input the hostname:localhost
Do you want to add hosts:[y or n,default no]:y
plz enter hostname in hosts:baicai
plz enter hostname ip in hosts:192.168.100.11
Do you want to add hosts:[y or n,default no]:
127.0.0.1    localhost localhost.localdomain localhost4 localhost4.localdomain4
::1          localhost localhost.localdomain localhost6 localhost6.localdomain6
 localhost
192.168.100.11 baicai
```

Shell 脚本运行截图如图 1-1 所示。

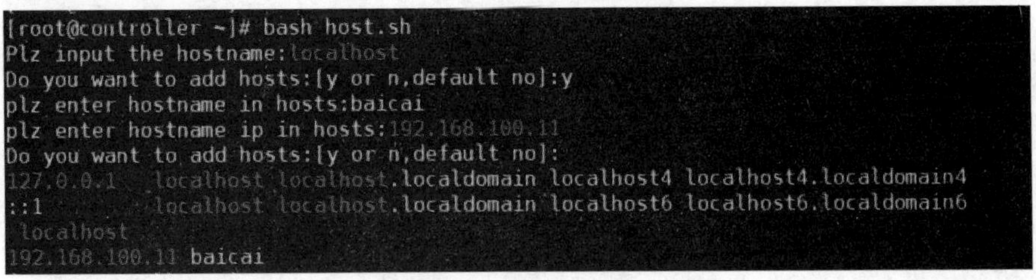

图 1-1 运行结果

本任务到此结束。

1.4 开放研究任务：Shell 脚本项目实施

掌握 Shell 编程是每一个 Linux 系统管理员必备的技能，把需要计算机执行的那些 Linux 命令罗列到一个文件里，再加上一些控制语句，这就是一个 Shell 程序，而且不用编译，Shell 程序是一种解释语言，执行 Shell 程序有两种方法，一个是"bash shell/sh shell 程序名"，另一种方法是先赋予其可执行权限，然后再执行即可。

任务执行清单

在本任务中，您将通过所学 Shell 知识，使用 Shell 脚本技术编写 IP、主机、yum 源的配置脚本文件以及编写菜单模块功能。

（1）Shell 脚本之 IP 配置。
（2）Shell 脚本之主机配置。
（3）Shell 脚本之 yum 配置。
（4）Shell 脚本之菜单栏。

目标

- 掌握变量的使用。
- 掌握算术运算的使用。
- 掌握循环语句的使用。
- 掌握正则表达式的使用。
- 掌握 Shell 脚本编写。

重要信息

- 本任务采用 CentOS 7 操作系统。
- 虚拟机密码为 000000。

解决方案

1. 编写 Shell 脚本

```
[root@localhost ~]# cat example.sh
#!/bin/bash
# Date:2021-2-15
# Version:0.1
# This program is system test
# Author:baicai
LANG=C;export LANG
menu()
{
clear
author="baicai"
host=`hostname`
softversion=1.0.0
revision=20210215
date=`date +%F.%T`
# less echo,faster speed

cat << MENULIST
############################ 菜单栏配置 ############################
Author:$author Host:$host Version:$softversion Revision:$revision
--------------------------------------------------------------------
This shell script can do these:
configure
1.configure IP-ADDRESS/BROADCAST/NETMASK
2.configure hostname
3.configure localrepo
```

```
----------------------------------------------
check
4.check IP-ADDRESS
5.check hostname
6.check localrepo
7.check all
##################################################################
MENULIST
echo -n "plz enter u choice [1,2,3,4,5,6,7,b(back),q(quit)]:"
read choice
}
#######################IP 配置 ####################################
c_netinfo(){
CARDNUM=0
MODENUM=0
CARDNAME=ens33
BOOTPT="dhcp"

while[[ $CARDNUM != 1 ]] && [[ $CARDNUM != 2]]
do
CARDNUM=1
echo -n "Plz enter the network device:[1(ens33),2(ens34),default 1]:"
read CARDNUM
if[[ $CARDNUM == 1]];
then
        CARDNAME=ens33
        break
elif[[ $CARDNUM == 2]];
then
        CARDNAME=ens34
        CARDNUM=1
        break
elif[[$CARDNUM == ""]];
then
        CARDNAME=ens33
        break
else
        echo "Plz input choice 1 or 2[1(ens33),2(ens34),default 1]:"
fi
done

cat > /tmp/ifcfg-iptmp << EOF
DEVICE=$CARDNAME
BOOTPROTO=$BOOTPT
ONBOOT="yes"
TYPE="Ethernet"
USERCTL="yes"
PEERDNS="yes"
IPV6INIT="no"
PERSISTENT_DHCLIENT="1"
EOF
####################### 网络模式配置 ############################
while[[ $MODENUM != 1]] && [[$MODENUM != 2]]
do
MODENUM=1
```

```
    echo -n "Plz enter the network mode:[1(dhcp),2(static),default 1]:"
    read MODENUM

    if[[ $MODENUM == 1]]||[[$MODENUM == ""]];
    then
            cp /tmp/ifcfg-iptmp /etc/sysconfig/network-scripts/ifcfg-$CARDNAME
            systemctl restart network
            break
    elif[[$MODENUM == 2]];
    then
        sed -i 's/dhcp/static/g' /tmp/ifcfg-iptmp
            echo -n "plz enter the IPADDRESS:"
        read IPADDR
            echo IPADDR=$IPADDR >> /tmp/ifcfg-iptmp
        echo -n "plz enter the NETMASK:"
        read NETMASK
            echo NETMASK=$NETMASK >> /tmp/ifcfg-iptmp
######################### 网关配置 ##############################
    echo -n "Do you want to configure the GATEWAY(y,n,default(n),enter to no configure)):"
        read GWCHOICE
        [[ $GWCHOICE == "" ]]&&GWCHOICE="n"
    case "$GWCHOICE" in
    "y")
        echo -n "plz enter the GATEWAY:"
        read GATEWAY
        echo GATEWAY=$GATEWAY >> /tmp/ifcfg-iptmp
    ;;
    "n")
    ;;
    *)
        echo -n "Inpute Error."
    ;;
    esac
############################DNS 配置 ##############################
    echo -n "Do you want to configure the DNS(y,n,default(none),enter to no configure)):"
    read DNSCHOICE
    [[ $DNSCHOICE == "" ]] && DNSCHOICE="n"
    case "$DNSCHOICE" in
    "y")
        echo -n "plz enter the DNS:"
        read DNS
            echo DNS=$DNS >> /tmp/ifcfg-iptmp
    ;;
    "n")
    ;;
    *)
    ;;
    esac

    cp /tmp/ifcfg-iptmp /etc/sysconfig/network-scripts/ifcfg-$CARDNAME
    systemctl restart network

    else
```

```bash
            echo -n "Plz input choice 1 or 2,enter to reinstall..."
            read MODENUM
    fi
done

}
########################### 主机配置 ##################################
c_hostname(){
hostname=""
HOSTNUM="y"
echo -n "Plz input the hostname:"
read hostname
hostnamectl set-hostname $hostname
sed -i '3,$d' /etc/hosts
echo "$IPADDR $hostname" >> /etc/hosts
#while[[ $HOSTNUM != n ]] || [[ $HOSTNUM == "" ]]
while[[ True ]]
do
echo -n "Do you want to add hosts:[y or n,default no]:"
read HOSTNUM
[[ $HOSTNUM == "" ]] && HOSTNUM="n"
case "$HOSTNUM" in
"y")
    echo -n "plz enter hostname in hosts:"
    read hostname
    echo -n "plz enter hostname ip in hosts:"
    read IPADDR1
    echo "$IPADDR1 $hostname" >> /etc/hosts
;;
"n")
break
;;
*)
;;
esac
done

}
########################### YUM 配置 ##################################
c_repo(){
    [ ! -d /opt/centos7 ] && mkdir -p /opt/centos7
    [ ! -d /opt/bak ] && mkdir -p /opt/bak
    cat /etc/yum.repos.d/CentOS-* 2>1 1> /dev/null && mv -f /etc/yum.repos.d/CentOS-* /opt/bak
    echo -n "Which repo do you want to configure?[c(cdrom),l(local),default(cdrom):"
    read REPONUM
    [[ $REPONUM == "" ]] && REPONUM="n"
    case "$REPONUM" in
    "c")
    umount /opt/centos7
    mount /dev/cdrom /opt/centos7 || echo -n "mount cdrom fail."
    rm -rf /etc/yum.repos.d/*
```

```
        cat > /etc/yum.repos.d/cdrom.repo << EOF
        [centos7]
        name=centos7
        baseurl=file:///opt/centos7
        enabled=1
        gpgcheck=0
        EOF
        yum clean all
        yum repolist &&  echo "cdromrepo config successful." || echo "cdromrepo
config fail."
        ;;
        "l")
        echo -n "Please input the iso file path:"
        read FILEPATH
        [ ! -f $FILEPATH ] && echo "$FILEPATH not exsit."
        umount /opt/centos7
        mount -o loop $FILEPATH /opt/centos7
        rm -rf /etc/yum.repos.d/*
        cat > /etc/yum.repos.d/local.repo << EOF
        [centos7]
        name=centos7
        baseurl=file:///opt/centos7
        enabled=1
        gpgcheck=0
        EOF
        yum clean all
        yum repolist &&  echo "localrepo config successful." || echo "localrepo
config fail."
        ;;
        esac
        }
############################# 检查网络 ################################
check_net(){
MY_IP1=`ip a show dev ens33 | grep -w inet | awk '{print $2}'`
MY_IP2=`ip a show dev ens34 | grep -w inet | awk '{print $2}'`
GW=`route -n | grep UG | awk '{print $2}'`
DNS=`cat /etc/resolv.conf | grep name | awk '{print $2}'`
        echo "--------------NET IP INFORMATION--------------------"
        echo "ens33:   IP:$MY_IP1 "
        echo "ens34:   IP:$MY_IP2 "
        echo "GATEWAY:$GW "
        echo "DNS: $DNS "
        echo ""
        }
############################# 检查主机 ################################
check_hostname(){
H1=`hostname`
        echo "--------HOSTNAME AND /etc/hosts INFORMATION----------"
        echo "HOSTNAME=$H1"
        echo "/etc/hosts:"
        cat /etc/hosts
        echo ""
        }
```

```shell
########################## 检查YUM###################################
check_repo(){
echo "-----------------REPO----------------------"
yum repolist
echo ""
}
########################## 检查所有 ###################################
check_all(){
check_net
check_hostname
check_repo
}
###################################################################
for ((i=1;;i++))
do
menu
case "$choice" in
"1")
c_netinfo
;;
"2")
c_hostname
;;
"3")
c_repo
;;
"4")
check_net
;;
"5")
check_hostname
;;
"6")
check_repo
;;
"7")
check_all
;;
"q")
exit 0
;;
esac
if[ !"$choice"="" ]
then
echo "press any key to return!"
read
fi
done
```

2. 使用脚本配置 IP

在 Shell 脚本中已经将所有需要运行的命令使用函数包装了起来，只需要在想运行该程序的时候，输入相应的数值即可调用，图 1-2 所示为 Shell 脚本配置 IP 的运行流程结果。

```
#########################菜单栏配置####################################
Author:baicai Host:localhost Version:1.0.0 Revision:20210215
----------------------------------------------------------------------
This shell script can do these:
configure
1.configure IP-ADDRESS/BROADCAST/NETMASK
2.configure hostname
3.configure localrepo
--------------------------------------------
check
4.check IP-ADDRESS
5.check hostname
6.check localrepo
7.check all
###########################################################################
plz enter u choice [1,2,3,4,5,6,7,b(back),q(quit)]:1
Plz enter the network device:[1(ens33),2(ens34),default 1]:1
Plz enter the network mode:[1(dhcp),2(static),default 1]:2
plz enter the IPADDRESS:192.168.100.11
plz enter the NETMASK:255.255.255.0
Do you want to configure the GATEWAY(y,n,default(n),enter to no configure)):y
plz enter the GATEWAY:192.168.100.1
Do you want to configure the DNS(y,n,default(none),enter to no configure)):
press any key to return!
```

图 1-2　运行结果（1）

3. 使用脚本配置主机

Shell 脚本运行结果如图 1-3 所示。

```
#########################菜单栏配置####################################
Author:baicai Host:localhost Version:1.0.0 Revision:20210215
----------------------------------------------------------------------
This shell script can do these:
configure
1.configure IP-ADDRESS/BROADCAST/NETMASK
2.configure hostname
3.configure localrepo
--------------------------------------------
check
4.check IP-ADDRESS
5.check hostname
6.check localrepo
7.check all
###########################################################################
plz enter u choice [1,2,3,4,5,6,7,b(back),q(quit)]:2
Plz input the hostname:localhost
Do you want to add hosts:[y or n,default no]:y
plz enter hostname in hosts:baicai
plz enter hostname ip in hosts:192.168.100.11
Do you want to add hosts:[y or n,default no]:n
press any key to return!
```

图 1-3　运行结果（2）

4. 使用脚本配置 yum 源

在 yum 源文件脚本编写时，用到了比较运算符以及输出重定向等功能，Shell 脚本运行结果如图 1-4 所示。

```
#########################菜单栏配置####################################
Author:baicai Host:localhost Version:1.0.0 Revision:20210215
----------------------------------------------------------------------
This shell script can do these:
configure
1.configure IP-ADDRESS/BROADCAST/NETMASK
2.configure hostname
3.configure localrepo
--------------------------------------------
check
4.check IP-ADDRESS
5.check hostname
6.check localrepo
7.check all
###########################################################################
plz enter u choice [1,2,3,4,5,6,7,b(back),q(quit)]:3
Which repo do you want to configure?[c(cdrom),l(local),default(cdrom):c
umount: /opt/centos7: not mounted
mount: /dev/sr0 is write-protected, mounting read-only
Loaded plugins: fastestmirror
Cleaning repos: centos7
Cleaning up everything
Maybe you want: rm -rf /var/cache/yum, to also free up space taken by orphaned data from disabled or removed repos
Cleaning up list of fastest mirrors
Loaded plugins: fastestmirror
Determining fastest mirrors
centos7                                                          | 3.6 kB  00:00:00
(1/2): centos7/group_gz                                          | 166 kB  00:00:00
(2/2): centos7/primary_db                                        | 3.1 MB  00:00:00
repo id                               repo name                            status
centos7                               centos7                              3971
repolist: 3971
cdromrepo config successful.
press any key to return!
```

图 1-4　运行结果（3）

5. 使用脚本查看所有信息

如上所示，已基本根据任务要求完成 Shell 脚本的编写工作，利用 Shell 程序，可以将需要重复执行的若干命令组合在一起，通过执行程序，可以将组合在一起的所有命令一次性执行完毕，从而减少了重复地输入每一条命令的时间。

图 1-5 所示为 Shell 脚本的配置流程结果一览图，调用相应函数对应的值即可进行一键查看。

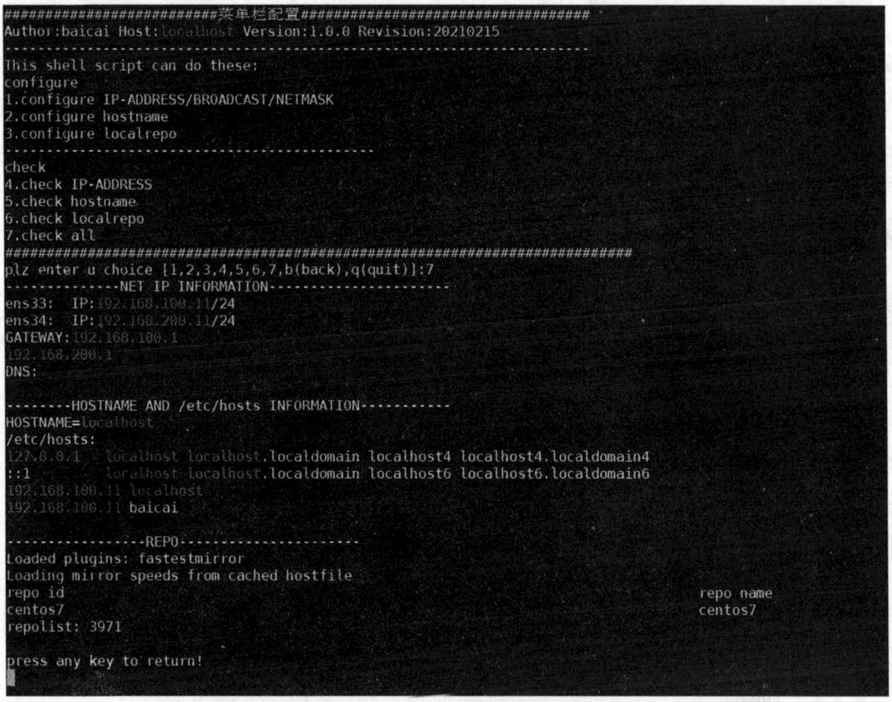

图 1-5　运行结果（4）

本任务到此结束。

小　　结

在本章中，您已经学会：
- 使用 Linux 操作系统。
- 使用 Linux 运维命令。
- 使用 vim 文件编辑器。
- 编写 Shell 脚本文件。

第 2 章 LAMP 与 LNMP

本章概要

学习 LAMP 与 LNMP 服务架构，部署 WordPress 个人博客网站。

学习目标

- 了解 LAMP 服务的原理和架构。
- 了解 LNMP 服务的原理和架构。
- 掌握 LAMP 服务架构部署 WordPress 博客网站。
- 掌握 LNMP 服务架构部署 WordPress 博客网站。

思维导图

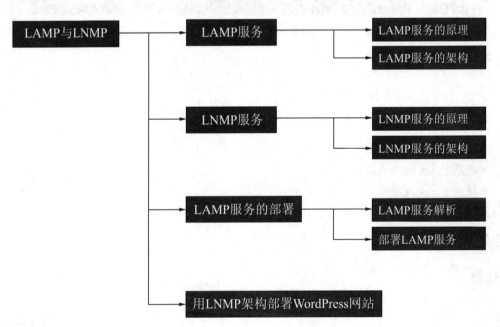

任务目标

LAMP 与 LNMP 服务架构部署 WordPress 个人博客网站。

2.1 LAMP 服务

学习目标

学完本节后，您应能够：
- 了解 LAMP 服务的原理。
- 了解 LAMP 服务的架构。

2.1.1 LAMP 服务的原理

1. LAMP 简介

LAMP 其实是指 Linux+Apache+MySQL+PHP 的结构体系，如图 2-1 所示。

图 2-1　LAMP 体系

- L：很显然 L 代表 Linux 系统，但此 L 需注意系统的版本号，如 CentOS 6.9 或 CentOS 7.2 等。
- A：表示 Apache，在传统行业中，多数采用 Apache 服务器，因此也很有必要了解学习 Apache。
- M：表示数据库，多数采用 MySQL 或 MariaDB，作为专业的数据库工程师需经多年的历练。
- P：表示 PHP、Python、Perl 等编程语言。

2. LAMP 服务原理

浏览器向服务器发送 HTTP 请求，服务器（Apache）接受请求，由于 PHP 作为 Apache 的组件模块也会一起启动，它们具有相同的生命周期。Apache 会将一些静态资源保存，然后去调用 PHP 处理模块进行 PHP 脚本的处理。脚本处理完后，Apache 将处理完的信息通过 http response 的方式发送给浏览器，浏览器经过解析、渲染等一系列操作后呈现整个网页。

2.1.2 LAMP 服务的架构

现今打开浏览器，搜索 LAMP 关键词，会出现大量关于 LAMP 的介绍，包括 LAMP 的一键脚本、LAMP 的 yum 安装、LAMP 的编译安装。但是对于一个非开发或非专业人员，有可能根据网络参考资源实现 LAMP 的搭建并成功运行各种服务，也有部分人员完全照搬某些博客知识进行搭建，最后以失败告终，因此抱怨互联网资源不够成熟，其实根本原因并非如此，其主要原因如下：

（1）初学者对 LAMP 架构原理不熟悉。
（2）初学者实验时所用系统和软件版本和某些博客资源并不相同。

因此大量初学者以失败告终，其实只有了解并掌握了 LAMP 的工作原理才能轻松地搭建 LAMP，其次参数的配置都是次要因素，因为互联网上拥有大量的参考资料可供查询。

在这里主要讲解 Linux+Apache+MariaDB+PHP 组合的架构，如图 2-2 所示。

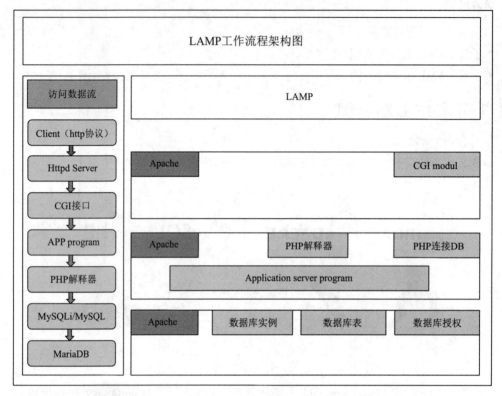

图 2-2　LAMP 架构

根据上图中访问数据流可知，处理一次动态页面请求，服务器主要经历：Apache 处理请求—通过 CGI 接口访问 PHP 的应用程序—PHP 应用程序调用 PHP 解释器执行 PHP 代码—PHP 程序访问调用数据库—最后反馈给客户。

在 LAMP 的环境机构中，Apache、MariaDB 和 PHP 的主要功能分别如下。

（1）Apache 主要实现如下功能：

① 处理 HTTP 的请求、构建响应报文等自身服务。

② 配置让 Apache 支持 PHP 程序的响应（通过 PHP 模块或 FPM）。

③ 配置 Apache 具体处理 PHP 程序的方法，如通过反向代理将 PHP 程序交给 fcgi 处理。

（2）MariaDB 主要实现如下功能：

① 提供 PHP 程序对数据的存储。

② 提供 PHP 程序对数据的读取（通常情况下从性能的角度考虑，尽量实现数据库的读写分离）。

（3）PHP 主要实现如下功能：

① 提供 Apache 的访问接口，即 CGI 或 Fast CGI（FPM）。

② 提供 PHP 程序的解释器。

③ 提供 MairaDB 数据库的连接函数的基本环境。

2.2 LNMP 服务

学习目标

学完本节后，您应能够：
- 了解 LNMP 服务的原理。
- 了解 LNMP 服务的架构。

2.2.1 LNMP 服务的原理

1. LNMP 简介

LNMP 是 Linux 系统下 Nginx+MySQL+PHP 的网站服务器架构。

2. LNMP 服务原理

首先 Nginx 服务不能处理动态请求，那么当用户发起动态请求时，Nginx 又是如何进行处理的？当用户发起 HTTP 请求，请求会被 Nginx 处理。如果是静态资源请求，Nginx 会直接返回；如果是动态请求，Nginx 则通过 fastcgi 协议转交给后端的 PHP 程序处理，具体如图 2-3 所示。

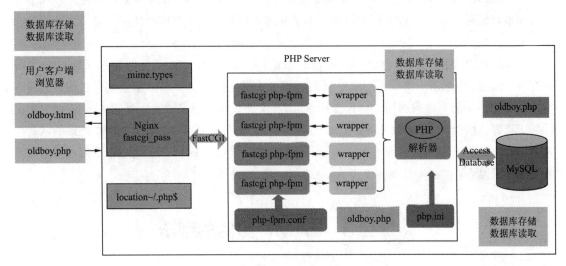

图 2-3　LNMP 原理

（1）用户通过 HTTP 协议发起请求，请求会先抵达 LNMP 架构中的 Nginx。

（2）Nginx 会根据用户的请求进行判断，这个判断由 Location 完成。

（3）判断用户请求的是静态页面，Nginx 直接进行处理。

（4）判断用户请求的是动态页面，Nginx 会将该请求交给 fastcgi 协议下发。

（5）fastcgi 会将请求交给 php-fpm 管理进程，php-fpm 管理进程接收到后会调用具体的工作进程 warrap。

（6）warrap 进程会调用 PHP 程序进行解析，如果只是解析代码，PHP 直接返回。

（7）如果有查询数据库操作，则由 PHP 连接数据库（用户、密码和 IP）发起查询的操作。

（8）最终数据由 MySQL->PHP->php-fpm->fastcgi->nginx->http->user。

2.2.2 LNMP 服务的架构

1. LNMP 服务架构

LNMP 是基于 CentOS/Debian 编写的 Nginx、PHP、MySQL、PHPMyAdmin、eAccelerator 一键安装包，可以在 VPS、独立主机上轻松地安装 LNMP 生产环境。

LNMP 代表的就是：Linux 系统下 Nginx+MySQL+PHP 这种网站服务器架构。

（1）Linux：是类 UNIX 计算机操作系统的统称，是目前最流行的免费操作系统。代表版本有：Debian、CentOS、Ubuntu、Fedora、Gentoo 等。

（2）Nginx：是一个高性能的 HTTP 和反向代理服务器，也是一个 IMAP/POP3/SMTP 代理服务器。

（3）MySQL：是一个小型关系型数据库管理系统。

（4）PHP：是一种在服务器端执行的嵌入 HTML 文档的脚本语言。

2. LNMP 和 LAMP 的区别

（1）LNMP 使用的是 Nginx，Nginx 是由 Igor Sysoev 为俄罗斯访问量第二的 Rambler.ru 站点开发的，第一个公开版本 0.1.0 发布于 2004 年 10 月 4 日，2011 年 6 月 1 日，nginx 1.0.4 发布。

（2）LAMP 使用的是 Apache，Apache 是世界实用排名第一的 Web 服务器软件，其几乎可以在所有广泛使用的计算机平台上运营，由于其跨平台和安全性被广泛使用，是最流行的 Web 服务端软件之一。

（3）LNMP 架构里 PHP 会启动服务 php-fpm，而 LAMP 中 php 只是作为 Apache 的一个模块存在。Nginx 会把用户的动态请求交给 PHP 服务去做处理，这个 PHP 服务就会去和数据库进行交互。用户的静态请求 Nginx 会直接处理，Nginx 处理静态请求的速度要比 Apache 快很多，性能上要好，所以 Apache 和 Nginx 在动态请求处理上区别不大，但是如果是静态处理的话就会发现 Nginx 要快于 Apache，而且 Nginx 能承受的并发量要比 Apache 大，可以承受好几万的并发连接量，所以大一些的网站都会使用 Nginx 作为 Web 服务器。

2.3 LAMP 服务的部署

学习目标

学完本节后，您应能够：

- 掌握 LAMP 服务的安装。
- 掌握 LAMP 服务的配置。
- 掌握 LAMP 架构搭建 WordPress 网站。

2.3.1 LAMP 服务解析

1. Apache 服务

Apache 是一种开放源码的 HTTP 服务器，可以在大多数计算机操作系统中运行，由于其多平台和安全性被广泛使用，是最流行的 Web 服务器端软件之一。它快速、可靠，并且可通过简单的 API 扩展，Perl/Python 等解释器可被编译到服务器中。

2. MySQL 服务

MySQL 是一个关系型数据库管理系统，关系数据库将数据保存在不同的表中，而不是将所有数据放在一个大仓库内，这样就增加了速度并提高了灵活性。MySQL 所使用的 SQL 语言是用于访问数据库最常用的标准化语言。

MySQL 逻辑架构整体分为三层，从上到下依次为客户层、核心服务层、存储引擎层。如图 2-4 所示为 MySQL 数据库架构图。

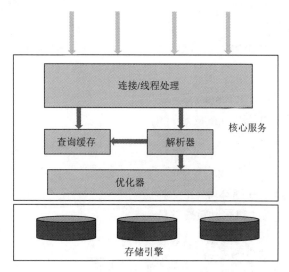

图 2-4　MySQL 架构图

（1）客户层。最上层是一些客户端和连接服务，包含本地的 sock 通信和大多数基于客户端/服务端工具实现的类似于 TCP/IP 的通信，主要完成一些类似于连接处理、授权认证及相关的安全方案，在该层上引用了线程池的概念，为通过认证安全接入的客户端提供线程。同样在该层上可以实现基于 SSL 的安全连接。服务器也会为安全接入的每个客户端验证它所具有的操作权限。

（2）核心服务层。第二层架构主要完成大多数的核心服务功能。如 SQL 接口，并完成缓存的查询。SQL 的分析和优化以及部分内置函数的执行。所有跨存储引擎的功能也在这一层实现，如过程、函数等。在该层，服务器会解析查询并创建相应的内部解析树，并对其完成相应的优化如确定查询表的顺序，是否利用索引等。最后生成相应的执行操作。

（3）存储引擎层。存储引擎负责 MySQL 中数据的存储和提取，服务器通过 API 与存储引擎进行通信，不同的存储引擎具有的功能不同，这样可以根据实际需要进行选取。

3. PHP 服务

PHP 即"超文本预处理器"脚本语言。PHP 是在服务器端执行的脚本语言，与 C 语言类似，是常用的网站编程语言。PHP 独特的语法混合了 C、Java、Perl 以及 PHP 自创的语法，利于学习，使用广泛，主要适用于 Web 开发领域。

2.3.2　任务：部署 LAMP 服务

WordPress 是一个注重美学、易用性和网络标准的个人信息发布平台。WordPress 虽为免费的开源软件，但其价值无法用金钱来衡量。使用 WordPress 可以构建功能强大的网络信息发布平台，但更多的是应用于个性化的博客。针对博客的应用，WordPress 能省略后台复杂的代码，让用户集中精力做好网站的内容。

📋 任务执行清单

在本任务中，您将通过实践掌握 LAMP 服务安装和配置，学会使用 LAMP 服务架构，部署 WordPress 服务。

☕ 目标

- 掌握 LAMP 服务的安装。
- 掌握 LAMP 服务的配置。
- 掌握 LAMP 架构搭建 WordPress 网站。

📎 重要信息

- 镜像的主要挂载点是 /opt/centos。
- 此次实验所有服务均采用 yum 安装。
- 虚拟机密码为 000000。

✏️ 解决方案

1. 登录虚拟机，配置基础环境

（1）登录虚拟机，更改主机名并重新登录。

```
[root@localhost ~]# hostnamectl set-hostname LAMP
[root@localhost ~]# logout
[root@lnmp ~]#
```

（2）在 lamp 节点上创建 /opt/centos 挂载点。

```
[root@lamp ~]# mkdir /opt/centos
```

（3）在 lamp 节点上，使用 mount 命令挂载 CentOS 7 镜像，并使用命令验证是否正确挂载到 /opt/centos 挂载点上。

```
[root@lamp ~]# mount -o loop CentOS-7.4-x86_64-DVD-1804.iso  /opt/centos/
mount: /dev/loop0 is write-protected, mounting read-only
[root@lamp ~]# df -h
Filesystem         Size    Used  Avail Use%  Mounted on
/dev/vda1           30G    5.2G   25G   18%  /
devtmpfs           898M       0  898M    0%  /dev
tmpfs              920M       0  920M    0%  /dev/shm
tmpfs              920M     17M  903M    2%  /run
tmpfs              920M       0  920M    0%  /sys/fs/cgroup
tmpfs              184M       0  184M    0%  /run/user/0
/dev/loop0         4.2G    4.2G     0  100%  /opt/centos
```

（4）列出已挂载的镜像内容。

```
[root@lamp ~]# ls -l /opt/centos/
total 678
-rw-rw-r-- 1 root root         14 May  2  2018 CentOS_BuildTag
drwxr-xr-x 3 root root       2048 May  3  2018 EFI
-rw-rw-r-- 1 root root        227 Aug 30  2017 EULA
-rw-rw-r-- 1 root root      18009 Dec  9  2015 GPL
drwxr-xr-x 3 root root       2048 May  3  2018 images
drwxr-xr-x 2 root root       2048 May  3  2018 isolinux
drwxr-xr-x 2 root root       2048 May  3  2018 LiveOS
drwxrwxr-x 2 root root     655360 May  3  2018 Packages
drwxrwxr-x 2 root root       4096 May  3  2018 repodata
```

```
-rw-rw-r-- 1 root root   1690 Dec  9  2015 RPM-GPG-KEY-CentOS-7
-rw-rw-r-- 1 root root   1690 Dec  9  2015 RPM-GPG-KEY-CentOS-Testing-7
-r--r--r-- 1 root root   2883 May  3  2018 TRANS.TBL
```

（5）删除默认 /etc/yum.repos.d/ 目录下的 yum 源，手工创建 yum 源配置文件 local.repo。

```
[root@lamp ~]# rm -rf /etc/yum.repos.d/*
[root@lamp ~]# vi /etc/yum.repos.d/local.repo
[centos]
name=centos
baseurl=file:///opt/centos
enabled=1
gpgcheck=0
```

（6）使用命令清除缓存文件，并确认 yum 源是否配置成功。

```
[root@lamp ~]# yum clean all
Loaded plugins: fastestmirror
Cleaning repos: centos
Cleaning up everything
Maybe you want: rm -rf /var/cache/yum, to also free up space taken by
orphaned data from disabled or removed repos
[root@nginx ~]# yum repolist
Loaded plugins: fastestmirror
Determining fastest mirrors
centos                                           | 3.6 kB  00:00:00
(1/2): centos/group_gz                           | 166 kB  00:00:00
(2/2): centos/primary_db                         | 3.1 MB  00:00:00
repo id                   repo name                     status
centos                    centos                        3,971
repolist: 3,971
```

2. LAMP 服务安装

（1）安装 httpd、php、mariadb 服务。

```
[root@lamp ~]# yum install -y httpd httpd-devel mariadb-server mariadb-
devel php
 php-mysql php-devel
```

如图 2-5 所示则表明安装成功。

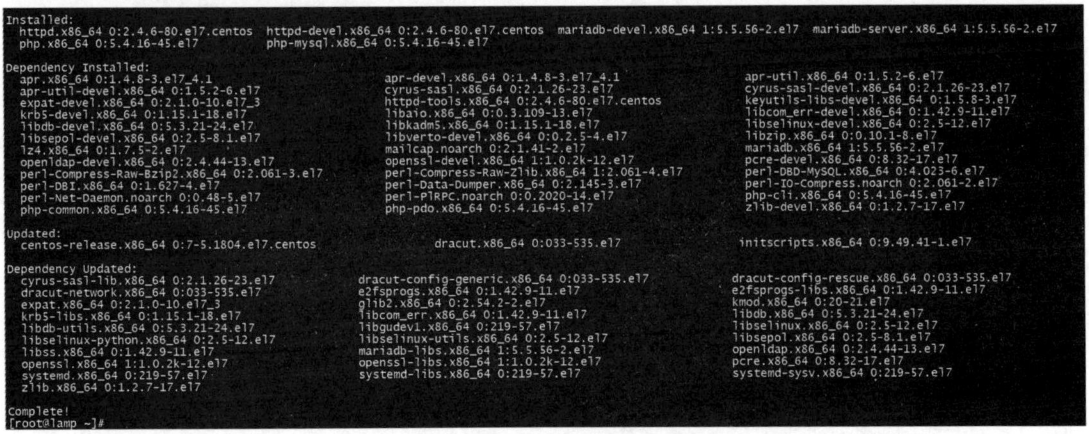

```
Installed:
  httpd.x86_64 0:2.4.6-80.el7.centos  httpd-devel.x86_64 0:2.4.6-80.el7.centos  mariadb-devel.x86_64 1:5.5.56-2.el7  mariadb-server.x86_64 1:5.5.56-2.el7
  php.x86_64 0:5.4.16-45.el7          php-mysql.x86_64 0:5.4.16-45.el7

Dependency Installed:
  apr.x86_64 0:1.4.8-3.el7_4.1                     apr-devel.x86_64 0:1.4.8-3.el7_4.1               apr-util.x86_64 0:1.5.2-6.el7
  apr-util-devel.x86_64 0:1.5.2-6.el7              cyrus-sasl.x86_64 0:2.1.26-23.el7                cyrus-sasl-devel.x86_64 0:2.1.26-23.el7
  expat-devel.x86_64 0:2.1.0-10.el7_3              httpd-tools.x86_64 0:2.4.6-80.el7.centos         keyutils-libs-devel.x86_64 0:1.5.8-3.el7
  krb5-devel.x86_64 0:1.15.1-18.el7                libaio.x86_64 0:0.3.109-13.el7                   libcom_err-devel.x86_64 0:1.42.9-11.el7
  libdb-devel.x86_64 0:5.3.21-24.el7               libkadm5.x86_64 0:1.15.1-18.el7                  libselinux-devel.x86_64 0:2.5-12.el7
  libsepol-devel.x86_64 0:2.5-8.1.el7              libverto-devel.x86_64 0:0.2.5-4.el7              libzip.x86_64 0:0.10.1-8.el7
  lz4.x86_64 0:1.7.5-2.el7                         mailcap.noarch 0:2.1.41-2.el7                    mariadb.x86_64 1:5.5.56-2.el7
  openldap-devel.x86_64 0:2.4.44-13.el7            openssl-devel.x86_64 1:1.0.2k-12.el7             pcre-devel.x86_64 0:8.32-17.el7
  perl-Compress-Raw-Bzip2.x86_64 0:2.061-3.el7     perl-Compress-Raw-Zlib.x86_64 1:2.061-4.el7      perl-DBD-MySQL.x86_64 0:4.023-6.el7
  perl-DBI.x86_64 0:1.627-4.el7                    perl-Data-Dumper.x86_64 0:2.145-3.el7            perl-IO-Compress.noarch 0:2.061-2.el7
  perl-Net-Daemon.noarch 0:0.48-5.el7              perl-PlRPC.noarch 0:0.2020-14.el7                php-cli.x86_64 0:5.4.16-45.el7
  php-common.x86_64 0:5.4.16-45.el7                php-pdo.x86_64 0:5.4.16-45.el7                   zlib-devel.x86_64 0:1.2.7-17.el7

Updated:
  centos-release.x86_64 0:7-5.1804.el7.centos          dracut.x86_64 0:033-535.el7          initscripts.x86_64 0:9.49.41-1.el7

Dependency Updated:
  cyrus-sasl-lib.x86_64 0:2.1.26-23.el7        dracut-config-generic.x86_64 0:033-535.el7     dracut-config-rescue.x86_64 0:033-535.el7
  dracut-network.x86_64 0:033-535.el7          e2fsprogs.x86_64 0:1.42.9-11.el7               e2fsprogs-libs.x86_64 0:1.42.9-11.el7
  expat.x86_64 0:2.1.0-10.el7_3                glib2.x86_64 0:2.54.2-2.el7                    kmod.x86_64 0:20-21.el7
  krb5-libs.x86_64 0:1.15.1-18.el7             libcom_err.x86_64 0:1.42.9-11.el7              libdb.x86_64 0:5.3.21-24.el7
  libdb-utils.x86_64 0:5.3.21-24.el7           libgudev1.x86_64 0:219-57.el7                  libselinux.x86_64 0:2.5-12.el7
  libselinux-python.x86_64 0:2.5-12.el7        libselinux-utils.x86_64 0:2.5-12.el7           libsepol.x86_64 0:2.5-8.1.el7
  libss.x86_64 0:1.42.9-11.el7                 mariadb-libs.x86_64 1:5.5.56-2.el7             openldap.x86_64 0:2.4.44-13.el7
  openssl.x86_64 1:1.0.2k-12.el7               openssl-libs.x86_64 1:1.0.2k-12.el7            pcre.x86_64 0:8.32-17.el7
  systemd.x86_64 0:219-57.el7                  systemd-libs.x86_64 0:219-57.el7               systemd-sysv.x86_64 0:219-57.el7
  zlib.x86_64 0:1.2.7-17.el7

Complete!
[root@lamp ~]#
```

图 2-5　安装服务

（2）重启服务 httpd 服务与 mariadb 服务，并设置开机自启。

```
[root@lamp ~]# systemctl restart {httpd,mariadb}
[root@lamp ~]# systemctl enable  {httpd,mariadb}
```

```
    Created symlink from /etc/systemd/system/multi-user.target.wants/httpd.
service to /usr/lib/systemd/system/httpd.service.
    Created symlink from /etc/systemd/system/multi-user.target.wants/mariadb.
service to /usr/lib/systemd/system/mariadb.service.
```

使用命令查看服务的状态，结果如图 2-6 所示。

图 2-6　服务状态

（3）创建数据库，设置用户以及密码。

```
[root@lamp ~]# mysqladmin -uroot password 000000
```

（4）使用用户和密码登入数据库。

```
[root@lamp ~]# mysql -uroot -p000000
Welcome to the MariaDB monitor.  Commands end with ; or \g.
Your MariaDB connection id is 3
Server version: 5.5.56-MariaDB MariaDB Server
Copyright (c) 2000, 2017, Oracle, MariaDB Corporation Ab and others.
Type 'help;' or '\h' for help. Type '\c' to clear the current input statement.
MariaDB [(none)]>
```

3. 测试部署

（1）系统界面测试，在浏览器中输入 IP 地址将看到图 2-7 所示界面。

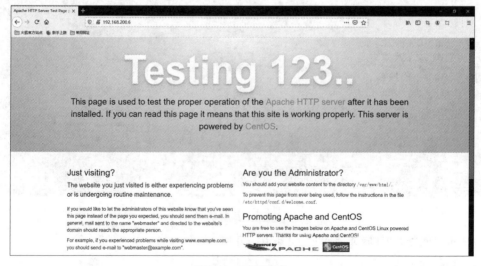

图 2-7　网页测试

（2）PHP 文件测试，将 httpd 的初始界面移走。

```
[root@lamp ~]# mv /etc/httpd/conf.d/welcome.conf
/etc/httpd/conf.d/welcome.conf.bak
```

（3）在 /var/www/html 下创建一个新的 index.php 文件。

```
[root@lamp ~]# cd /var/www/html/
[root@lamp html]# vi index.php
<?php
        phpinfo();
?>
```

PHP 文件测试成功的界面如图 2-8 所示。

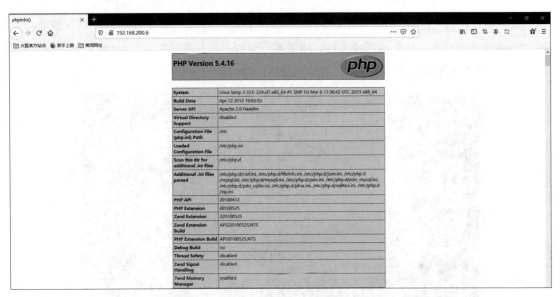

图 2-8　PHP 文件测试

4. 部署 WordPress 应用

（1）在部署 WordPress 之前，还需要做几个基础的配置，首先是数据库，需要登录数据库，创建 WordPress 数据库并赋予远程权限，命令如下所示：

```
[root@lamp ~]# mysql -uroot -p000000
Welcome to the MariaDB monitor.  Commands end with ; or \g.
Your MariaDB connection id is 4
Server version: 5.5.56-MariaDB MariaDB Server
Copyright (c) 2000, 2017, Oracle, MariaDB Corporation Ab and others.
Type 'help;' or '\h' for help. Type '\c' to clear the current input statement.
MariaDB [(none)]> create database wordpress;
Query OK, 1 row affected (0.00 sec)
MariaDB [(none)]> grant all privileges on *.* to root@'localhost' identified by '000000';
Query OK, 0 rows affected (0.00 sec)
MariaDB [(none)]> grant all privileges on *.* to root@'%' identified by '000000';
Query OK, 0 rows affected (0.00 sec)
MariaDB [(none)]> flush privileges;
Query OK, 0 rows affected (0.00 sec)
```

```
MariaDB [(none)]> exit
Bye
```

（2）将提供的 wordpress-4.7.3-zh_CN.zip 压缩包上传至虚拟机的 /root 目录并解压，命令如下：

```
[root@lamp ~]# yum -y install unzip
[root@lamp ~]# unzip wordpress-4.7.3-zh_CN.zip
```

（3）将 /root/wordpress 目录下的所有文件，复制到 /var/www/html/ 目录下，并赋予 777 的权限，命令如下所示：

```
[root@lamp ~]# cp -rvf wordpress/* /var/www/html/
[root@lamp ~]# chmod 777 -R /var/www/html/
```

（4）在 /var/www/html/ 目录下，可以看见一个 wp-config-sample.php 配置文件，该文件是 WordPress 应用提供的一个模板配置文件，将该模板复制一份并改名为 wp-config.php，然后编辑该文件，命令和结果如图 2-9 所示

```
[root@lamp ~]# cd /var/www/html/
[root@lamp html]# cp wp-config-sample.php wp-config.php
[root@lamp html]# vi wp-config.php
<?php
/**
 * WordPress基础配置文件。
 *
 * 这个文件被安装程序用于自动生成wp-config.php配置文件，
 * 您可以不使用网站，您需要手动复制这个文件，
 * 并重命名为“wp-config.php”，然后填入相关信息。
 *
 * 本文件包含以下配置选项：
 *
 * * MySQL设置
 * * 密钥
 * * 数据库表名前缀
 * * ABSPATH
 *
 * @link https://codex.wordpress.org/zh-cn:%E7%BC%96%E8%BE%91_wp-config.php
 *
 * @package WordPress
 */

// ** MySQL 设置 - 具体信息来自您正在使用的主机 ** //
/** WordPress数据库的名称 */
define('DB_NAME', 'wordpress');

/** MySQL数据库用户名 */
define('DB_USER', 'root');

/** MySQL数据库密码 */
define('DB_PASSWORD', '000000');

/** MySQL主机 */
define('DB_HOST', '127.0.0.1');

/** 创建数据表时默认的文字编码 */
define('DB_CHARSET', 'utf8');
```

图 2-9　WordPress 配置

（5）修改完毕后，保存退出，在浏览器中输入地址 192.168.200.6，刷新页面，进入 WordPress 安装界面，填写必要信息，单击“安装 WordPress”按钮，如图 2-10 所示。安装完毕后，刷新页面，单击“登录”按钮，如图 2-11 所示。

在登录界面输入安装时填写的账户和密码信息，单击“登录”按钮，如图 2-12 所示。登录后，进入 WordPress 应用的后台仪表盘界面，如图 2-13 所示。

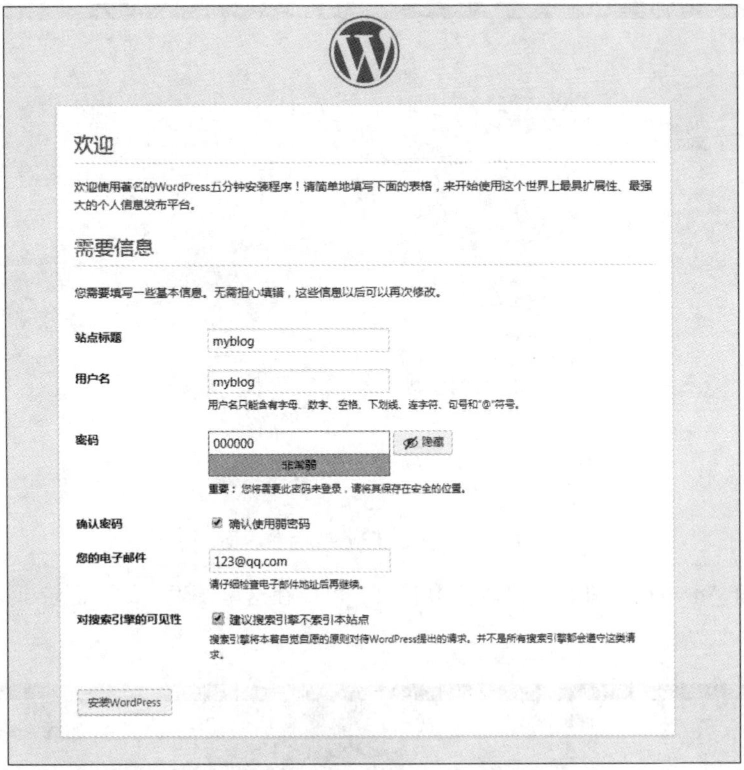

图 2-10　WordPress 安装界面

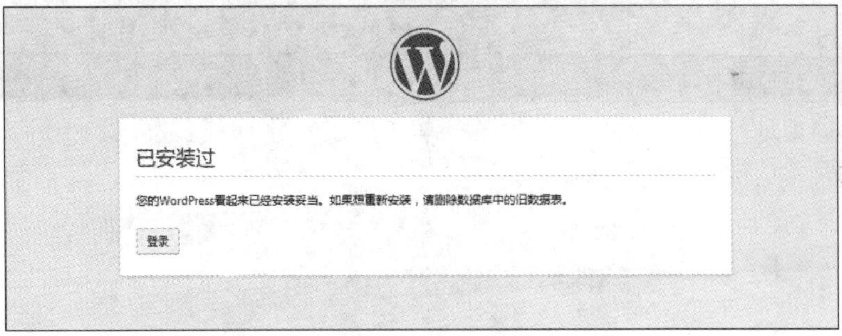

图 2-11　安装完毕

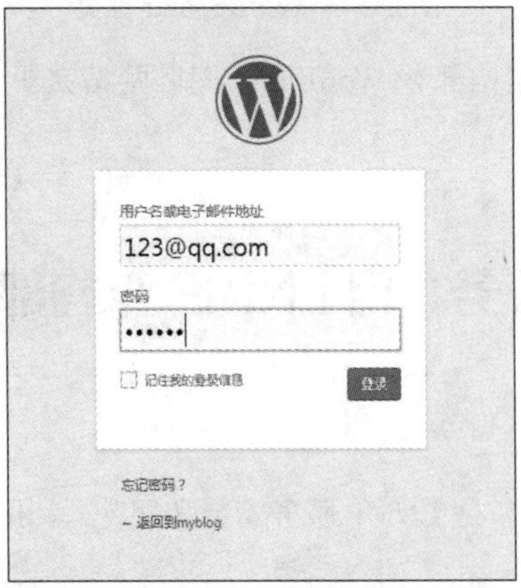

图 2-12　登录界面

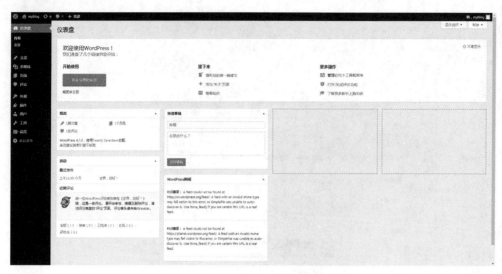

图 2-13　WordPress 后台界面

单击左上角"myblog"图标，进入博客首页，可以在这里发表文章，记录事迹等，如图 2-14 所示。

图 2-14　WordPress 首页

本任务到此结束，LAMP 架构部署 WordPress 博客网站完毕，学习动手搭建一个属于自己的博客系统吧。

2.4　开放研究任务：用 LNMP 架构部署 WordPress 网站

任务执行清单

在本任务中，您将通过实践掌握 LNMP 服务安装和配置，利用 LNMP 服务架构部署 WordPress 个人博客网站。

（1）配置基础环境。
（2）Nginx 服务配置。
（3）MySQL 服务配置。
（4）PHP 服务配置。
（5）WordPress 网站构建。

☕ **目标**

- 掌握 LNMP 服务的安装。
- 掌握 LNMP 服务的配置。
- 掌握 LNMP 架构搭建 WordPress 网站。

📎 **重要信息**

- 镜像的主要挂载点是 /opt/centos。
- 此次实验所有服务均采用编译安装方式。
- 虚拟机密码为 000000。

✏️ **解决方案**

1. 配置基础环境

（1）登录虚拟机，更改主机名并重新登录。

```
[root@localhost ~]# hostnamectl set-hostname LAMP
[root@localhost ~]# logout
[root@lnmp ~]# vi /etc/hosts
127.0.0.1    localhost localhost.localdomain localhost4 localhost4.localdomain4
::1          localhost localhost.localdomain localhost6 localhost6.localdomain6
192.168.200.10 lnmp
```

（2）在 lnmp 节点上创建 /opt/centos 挂载点。

```
[root@lnmp ~]# mkdir /opt/centos
```

（3）在 lnmp 节点上，使用 mount 命令挂载 Cent OS7 镜像，并使用命令验证是否正确挂载到 /opt/centos 挂载点上。

```
[root@lnmp ~]# mount -o loop CentOS-7.4-x86_64-DVD-1804.iso /opt/centos/
mount: /dev/loop0 is write-protected, mounting read-only
[root@lnmp ~]# df -h
Filesystem      Size  Used Avail Use% Mounted on
/dev/vda1        30G  5.2G   25G  18% /
devtmpfs        898M     0  898M   0% /dev
tmpfs           920M     0  920M   0% /dev/shm
tmpfs           920M   17M  903M   2% /run
tmpfs           920M     0  920M   0% /sys/fs/cgroup
tmpfs           184M     0  184M   0% /run/user/0
/dev/loop0      4.2G  4.2G     0 100% /opt/centos
```

（4）列出已挂载的镜像内容。

```
[root@lnmp ~]# ls -l /opt/centos/
total 678
-rw-rw-r-- 1 root root       14 May  2  2018 CentOS_BuildTag
```

```
drwxr-xr-x  3 root root    2048 May  3  2018 EFI
-rw-rw-r--  1 root root     227 Aug 30  2017 EULA
-rw-rw-r--  1 root root   18009 Dec  9  2015 GPL
drwxr-xr-x  3 root root    2048 May  3  2018 images
drwxr-xr-x  2 root root    2048 May  3  2018 isolinux
drwxr-xr-x  2 root root    2048 May  3  2018 LiveOS
drwxrwxr-x  2 root root  655360 May  3  2018 Packages
drwxrwxr-x  2 root root    4096 May  3  2018 repodata
-rw-rw-r--  1 root root    1690 Dec  9  2015 RPM-GPG-KEY-CentOS-7
-rw-rw-r--  1 root root    1690 Dec  9  2015 RPM-GPG-KEY-CentOS-Testing-7
-r--r--r--  1 root root    2883 May  3  2018 TRANS.TBL
```

（5）删除默认 /etc/yum.repos.d/ 目录下的 yum 源，手工创建 yum 源配置文件 local.repo。

```
[root@lnmp ~]# rm -rf /etc/yum.repos.d/*
[root@lnmp ~]# vi /etc/yum.repos.d/local.repo
[centos]
name=centos
baseurl=file:///opt/centos
enabled=1
gpgcheck=0
```

（6）使用命令清除缓存文件，并确认 yum 源是否配置成功。

```
[root@lnmp ~]# yum clean all
Loaded plugins: fastestmirror
Cleaning repos: centos
Cleaning up everything
Maybe you want: rm -rf /var/cache/yum, to also free up space taken by orphaned data from disabled or removed repos
[root@nginx ~]# yum repolist
Loaded plugins: fastestmirror
Determining fastest mirrors
centos                                                     | 3.6 kB  00:00:00
(1/2): centos/group_gz                                     | 166 kB  00:00:00
(2/2): centos/primary_db                                   | 3.1 MB  00:00:00
repo id                    repo name                                   status
centos                     centos                                       3,971
repolist: 3,971
```

2. Nginx 服务配置

（1）在安装之前首先创建指定用户，这里先创建 Nginx 用户，并检查是否创建成功，命令和结果如下所示：

```
[root@lnmp ~]# groupadd -g 1001 nginx
[root@lnmp ~]# useradd -u 900 nginx -g nginx -s /sbin/nologin
[root@lnmp ~]# tail -1 /etc/passwd
nginx:x:900:1001::/home/nginx:/sbin/nologin
```

（2）使用 yum 源安装依赖文件。

```
[root@lnmp ~]# yum -y install gcc gcc-c++ automake pcre pcre-devel zlib zlib-devel libxml2-devel libcurl-devel bzip2-devel openssl openssl-devel
```

（3）使用远程传输工具，将提供的 nginx-1.12.2.tar.gz 压缩包上传至 lnmp 节点的 /root/ 目录下，并解压到 /usr/local/src 目录中，命令和结果如下所示：

```
[root@lnmp ~]# tar -zxvf nginx-1.12.2.tar.gz -C /usr/local/src/
```

（4）编译安装，进入 nginx-1.12.2 目录，编译并安装，命令和结果如下所示：

```
[root@lnmp ~]# cd /usr/local/src/
[root@lnmp src]# cd nginx-1.12.2/
[root@lnmp nginx-1.12.2]# ./configure --prefix=/usr/local/nginx --with-http_dav_module \
--with-http_stub_status_module --with-http_addition_module \
--with-http_sub_module --with-http_flv_module --with-http_mp4_module \
--with-http_ssl_module --with-http_gzip_static_module --user=nginx --group=nginx
```

（5）如果没有报错提示，请进行下一步安装，命令如下所示：

```
[root@lnmp nginx-1.12.2]# make && make install
```

（6）编译安装完毕后，创建软连接并启动测试。

```
[root@lnmp nginx-1.12.2]#  ln -s /usr/local/nginx/sbin/nginx /usr/local/sbin/
[root@lnmp nginx-1.12.2]# nginx -t
nginx: the configuration file /usr/local/nginx/conf/nginx.conf syntax is ok
nginx: configuration file /usr/local/nginx/conf/nginx.conf test is successful
[root@lnmp ~]# nginx
```

（7）查看服务端口，如果发现 80 端口启动，则表示 Nginx 服务启动成功，如图 2-15 所示。

图 2-15　服务端口

> **注意：**
> - nginx -t 使用 -t 选项参数是用于检测文件语法是否有误。
> - 当 netstat 命令无法使用时，请自行使用 yum 源安装 net-tools 工具。

（8）在浏览器中访问地址 192.168.200.10，查看是否出现 Nginx 欢迎界面。结果如图 2-16 所示。

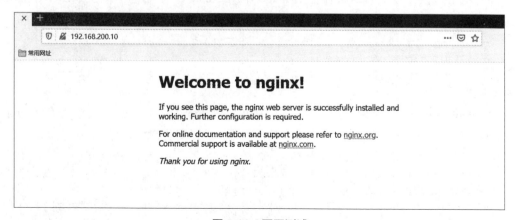

图 2-16　网页测试

3. MySQL 服务配置

（1）安装数据库服务，命令如下所示：

```
[root@lnmp ~]# yum install mariadb-server mariadb-devel
```

（2）启动数据库服务，创建数据库，设置用户以及密码，最后登录数据库进行验证，命令和结果如下所示：

```
[root@lnmp ~]# systemctl start mariadb
[root@lnmp ~]# mysqladmin -u root password 000000
[root@lnmp ~]# mysql -uroot -p000000
Welcome to the MariaDB monitor.  Commands end with ; or \g.
Your MariaDB connection id is 3
Server version: 5.5.56-MariaDB MariaDB Server
Copyright (c) 2000, 2017, Oracle, MariaDB Corporation Ab and others.
Type 'help;' or '\h' for help. Type '\c' to clear the current input statement.
MariaDB [(none)]> exit
Bye
```

（3）登录数据库，创建 WordPress 数据库并赋予远程权限，命令和结果如下所示：

```
[root@lnmp ~]# mysql -uroot -p000000
Welcome to the MariaDB monitor.  Commands end with ; or \g.
Your MariaDB connection id is 4
Server version: 5.5.56-MariaDB MariaDB Server
Copyright (c) 2000, 2017, Oracle, MariaDB Corporation Ab and others.
Type 'help;' or '\h' for help. Type '\c' to clear the current input statement.
MariaDB [(none)]> create database wordpress;
Query OK, 1 row affected (0.00 sec)
MariaDB [(none)]> grant all privileges on *.* to root@localhost identified by '000000' with grant option;
Query OK, 0 rows affected (0.00 sec)
MariaDB [(none)]> grant all privileges on *.* to root@"%" identified by '000000' with grant option;
Query OK, 0 rows affected (0.00 sec)
MariaDB [(none)]> exit
Bye
```

4. PHP 服务配置

（1）使用远程传输工具，将提供的 libmcrypt-2.5.8.tar.gz 压缩包上传至 PHP 节点的 /usr/local/src 目录下，解压该压缩包，进入解压后目录，编译安装该服务。

```
[root@lnmp ~]# tar -zxvf libmcrypt-2.5.8.tar.gz -C /usr/local/src/
[root@lnmp src]# cd libmcrypt-2.5.8/
[root@lnmp libmcrypt-2.5.8]# ./configure --prefix=/usr/local/libmcrypt && make && make install
```

（2）PHP 环境安装，使用远程传输工具，将提供的 php-5.6.27.tar.gz 压缩包上传至 lnmp 节点的 /root 目录下，解压该压缩包，进入解压后的目录，编译安装 PHP 服务。

```
[root@lnmp ~]# tar -zxvf php-5.6.27.tar.gz -C /usr/local/src/
[root@lnmp ~]# cd /usr/local/src/php-5.6.27/
[root@lnmp php-5.6.27]# ./configure --prefix=/usr/local/php5.6 --with-mysql=mysqlnd --with-pdo-mysql=mysqlnd --with-mysqli=mysqlnd \
--with-openssl --enable-fpm --enable-sockets --enable-sysvshm \
```

```
--enable-mbstring --with-freetype-dir --with-jpeg-dir --with-png-dir \
--with-zlib --with-libxml-dir=/usr --enable-xml --with-mhash \
--with-mcrypt=/usr/local/libmcrypt --with-config-file-path=/etc \
--with-config-file-scan-dir=/etc/php.d --with-bz2 --enable-maintainer-zts
```

（3）如果没有报错提示，则进行下一步的编译安装。

```
[root@php php-5.6.27]# make && make install
```

（4）PHP 环境配置，PHP 压缩包中提供了 PHP 环境需要用到的模板文件，需要对文件进行改名后才能使用，复制文件并改名。

```
[root@lnmp ~]# cd /usr/local/src/php-5.6.27/
[root@lnmp php-5.6.27]# cp php.ini-production  /etc/php.ini
[root@lnmp php-5.6.27]# cp sapi/fpm/init.d.php-fpm /etc/init.d/php-fpm
```

（5）因为安装服务需要用到执行权限，此处赋予文件的执行权限。

```
[root@lnmp php-5.6.27]# chmod +x /etc/init.d/php-fpm
```

（6）添加 PHP 服务到启动列表，并设置开机启动。

```
[root@lnmp php-5.6.27]# chkconfig --add php-fpm
[root@lnmp php-5.6.27]# chkconfig php-fpm on
```

（7）修改 PHP 的主配置文件 php-fpm.conf，并启动服务。

```
[root@lnmp php-5.6.27]# cp /usr/local/php5.6/etc/php-fpm.conf.default /usr/local/php5.6/etc/php-fpm.conf
[root@lnmp php-5.6.27]# vi /usr/local/php5.6/etc/php-fpm.conf
[root@lnmp php-5.6.27]# grep -n '^[a-Z] /usr/local/php5.6/etc/php-fpm.conf
149:user=nginx
150:group=nginx
164:listen=127.0.0.1:9000
224:pm=dynamic
235:pm.max_children=50
240:pm.start_servers=5
245:pm.min_spare_servers=5
250:pm.max_spare_servers=35
```

找到配置文件中的相应参数并修改，修改成上述配置。

（8）启动 PHP 服务，并检查是否启动成功，如果发现 9000 端口已启动，则说明 PHP 环境安装完毕。测试效果如图 2-17 所示。

```
[root@lnmp php-5.6.27]# systemctl restart php-fpm
[root@lnmp php-5.6.27]# netstat -ntapl
```

```
[root@lnmp php-5.6.27]# netstat -ntapl
Active Internet connections (servers and established)
Proto Recv-Q Send-Q Local Address           Foreign Address         State       PID/Program name
tcp        0      0 127.0.0.1:9000          0.0.0.0:*               LISTEN      9924/php-fpm: maste
tcp        0      0 0.0.0.0:3306            0.0.0.0:*               LISTEN      17185/mysqld
tcp        0      0 0.0.0.0:80              0.0.0.0:*               LISTEN      4414/nginx: master
tcp        0      0 0.0.0.0:22              0.0.0.0:*               LISTEN      948/sshd
tcp        0      0 127.0.0.1:25            0.0.0.0:*               LISTEN      867/master
tcp        0      0 10.0.0.7:22             172.16.180.225:50317    ESTABLISHED 16629/sshd: root@no
tcp        0      0 10.0.0.7:22             172.16.180.225:50316    ESTABLISHED 16627/sshd: root@pt
tcp        0      0 10.0.0.7:22             172.16.180.225:53559    ESTABLISHED 9480/sshd: root@not
tcp        0      0 10.0.0.7:22             172.16.180.225:53558    ESTABLISHED 9477/sshd: root@pts
tcp        0      0 10.0.0.7:22             172.16.180.225:51446    ESTABLISHED 30164/sshd: root@no
tcp        0     48 10.0.0.7:22             172.16.180.225:51445    ESTABLISHED 30044/sshd: root@pt
tcp6       0      0 :::22                   :::*                    LISTEN      948/sshd
tcp6       0      0 ::1:25                  :::*                    LISTEN      867/master
```

图 2-17　PHP 测试

5. WordPress 网站构建

（1）使用远程传输工具，将提供的 wordpress-4.7.3-zh_CN.zip 压缩包上传至 /root 目录下并解压。

```
[root@lnmp ~]# unzip wordpress-4.7.3-zh_CN.zip
```

🔔 **注意**：

如果出现 "-bash: unzip: command not found" 结果，表示命令没有安装，使用 yum 安装 unzip 命令即可解决。

（2）进入 /usr/local/nginx/html/ 目录，删除原有的 Nginx 网页文件，将解压好的 wordpress 配置文件复制到该目录下，并赋予最高权限，命令如下所示：

```
[root@lnmp ~]# cd /usr/local/nginx/html/
[root@lnmp html]# rm -rf index.html
[root@lnmp ~]# cp -avr wordpress/* /usr/share/nginx/html/
[root@lnmp wordpress]# cd /usr/local/nginx/html/
[root@lnmp html]# chmod 777 *
```

（3）在 /usr/local/nginx/html/ 目录下，可以看见一个 wp-config-sample.php 配置文件，该文件是 WordPress 应用提供了一个模板配置文件，将该模板复制一份并改名为 wp-config.php，然后编辑该文件，命令和结果如下所示：

```
[root@lnmp html]# cp wp-config-sample.php  wp-config.php
[root@lnmp html]# vi wp-config.php
[root@lnmp default]# vi wp-config.php
// ** MySQL 设置 - 具体信息来自您正在使用的主机 ** //
/** WordPress 数据库的名称 */
define('DB_NAME', 'wordpress');

/** MySQL 数据库用户名 */
define('DB_USER', 'root');

/** MySQL 数据库密码 */
define('DB_PASSWORD', '000000');

/** MySQL 主机 */
define('DB_HOST', '127.0.0.1');

/** 创建数据表时默认的文字编码 */
define('DB_CHARSET', 'utf8');

/** 数据库整理类型。如不确定请勿更改 */
define('DB_COLLATE', '');
```

（4）在 Nginx 配置文件中配置启动文件，配置完成后退出保存。

```
…省略…
    location / {
            root    html;
            index   index.php index.html index.htm;
    }
```

```
...省略...
location ~ \.php$ {
            root           html;
            fastcgi_pass   127.0.0.1:9000;
            fastcgi_index  index.php;
            fastcgi_param  SCRIPT_FILENAME  /scripts$fastcgi_script_name;
            include        fastcgi_params;
        }
```

(5) 接着在 /usr/local/nginx/conf/fastcgi_params 添加配置，命令如下所示：

```
[root@lnmp ~]# vi /usr/local/nginx/conf/fastcgi_params
fastcgi_param   QUERY_STRING        $query_string;
fastcgi_param   REQUEST_METHOD      $request_method;
fastcgi_param   CONTENT_TYPE        $content_type;
fastcgi_param   CONTENT_LENGTH      $content_length;

fastcgi_param   SCRIPT_NAME         $fastcgi_script_name;
fastcgi_param   SCRIPT_FILENAME     $document_root$fastcgi_script_name;
// 此处添加了一条指令，用于传递脚本文件请求的路径。SCRIPT_FILENAME 变量是 php-fpm 可识
别的脚本名称，对应的 $fastcgi_script_name 变量是 Nginx 定义的脚本名称，$document_root 变
量是 Nginx 配置的根目录，即 root 指定的路径
fastcgi_param   REQUEST_URI         $request_uri;
fastcgi_param   DOCUMENT_URI        $document_uri;
fastcgi_param   DOCUMENT_ROOT       $document_root;
fastcgi_param   SERVER_PROTOCOL     $server_protocol;
fastcgi_param   REQUEST_SCHEME      $scheme;
fastcgi_param   HTTPS               $https if_not_empty;

fastcgi_param   GATEWAY_INTERFACE   CGI/1.1;
fastcgi_param   SERVER_SOFTWARE     nginx/$nginx_version;

fastcgi_param   REMOTE_ADDR         $remote_addr;
fastcgi_param   REMOTE_PORT         $remote_port;
fastcgi_param   SERVER_ADDR         $server_addr;
fastcgi_param   SERVER_PORT         $server_port;
fastcgi_param   SERVER_NAME         $server_name;

# PHP only, required if PHP was built with --enable-force-cgi-redirect
fastcgi_param   REDIRECT_STATUS     200;
```

(6) 验证 WordPress 应用，重启 Nginx 服务，命令如下所示：

```
[root@lnmp ~]# nginx -s reload
```

(7) 在浏览器中输入 192.168.200.10 地址进行访问，会出现"著名的 WordPress 五分钟安装程序"，填写必要的信息，然后单击左下角"安装 WordPress"按钮，进行 WordPress 应用的安装，如图 2-18 所示。

图 2-18　WordPress 安装界面

（8）稍等片刻，安装完毕后，进入 WordPress 后台界面，如图 2-19 所示。

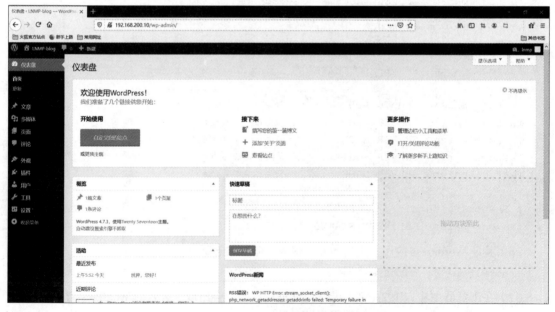

图 2-19　WordPress 后台界面

（9）可根据自定义配置相应的数据信息，并安装 WordPress 个人网站。单击图 2-19 页面左上角的"LNMP-blog"图标，进入 WordPress 首页，如图 2-20 所示。

图 2-20　WordPress 首页

本任务到此结束，LNMP 架构部署 WordPress 博客网站已完成。

小　　结

在本章中，您已经学会：
- LAMP 服务的原理和架构。
- LNMP 服务的原理和架构。
- 使用 LAMP 服务架构部署 WordPress 博客网站。
- 使用 LNMP 服务架构部署 WordPress 博客网站。

第 3 章

数据库运维与集群

本章概要

学习 MariaDB 数据库中主从复制与读写分离实验,并部署读写分离数据库集群。

学习目标

- 了解数据库服务的原理和架构。
- 掌握 MariaDB 数据库运维。
- 掌握 MariaDB 主从服务。
- 掌握 MariaDB 读写分离。
- 掌握 MariaDB 读写分离数据库集群。

思维导图

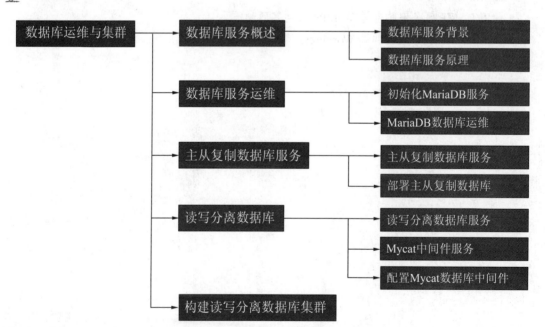

任务目标

使用 MariaDB 数据库中主从复制与读写分离实验部署读写分离数据库集群。

3.1 数据库服务概述

学习目标

学完本节后，您应能够：
- 了解数据库服务背景。
- 了解数据库服务原理。

3.1.1 数据库服务背景

1. MySQL 和 MariaDB

MySQL 在过去由于性能高、成本低、可靠性好，已经成为最流行的开源数据库，因此被广泛地应用在 Internet 上的中小型网站中。随着 MySQL 的不断成熟，它也逐渐用于更多大规模网站和应用，如阿里巴巴、腾讯、维基百科、Google 和 Facebook 等网站。非常流行的开源软件组合 LAMP 中的"M"指的就是 MySQL。

但被甲骨文公司收购后，Oracle 大幅调涨 MySQL 商业版的售价，且甲骨文公司不再支持另一个自由软件项目 OpenSolaris 的发展，因此导致自由软件社区们对于 Oracle 是否还会持续支持 MySQL 社区版（MySQL 之中唯一的免费版本）有所隐忧，MySQL 的创始人麦克尔·维德纽斯以 MySQL 为基础，成立分支计划 MariaDB。而原先一些使用 MySQL 的开源软件逐渐转向 MariaDB 或其他数据库。

MariaDB 数据库管理系统是 MySQL 的一个分支，主要由开源社区在维护，采用 GPL 授权许可。开发这个分支的原因之一是：甲骨文公司收购了 MySQL 后，有将 MySQL 闭源的潜在风险，因此社区采用分支的方式来避开这个风险。

2. 谷歌与 MariaDB

Google 为了顺利从 MySQL 迁移到 MariaDB，降低数据迁移成本，甚至派遣了一名工程师到 MariaDB 基金会协助开发，Google 和 MariaDB 基金会的合作关系已经显而易见。

2017 年，MariaDB 生态系统取得了很大的进展，包括开源、资金、协作、产品、服务。当然还包括开发和维护 MariaDB 代码本身。

2017 年，董事会选出了几名新成员。2017 年 6 月，董事会选出了 3 名新成员：阿里云的 Xiaobin Lin、IBM 的 Todd Boyd 和 Barry。12 月，又有两名新董事会成员当选：腾讯云的张青林和微软的 Tobias Ternstrom。

如图 3-1 所示的 MySQL 和 MariaDB，一个向左、一个向右。

图 3-1　MySQL 与 MariaDB

3.1.2 数据库服务原理

1. 数据库简介

数据库是数据管理的有效技术，是由一批数据构成的有序集合，这些数据被存放在结构化的数据表里。数据表之间相互关联，反映客观事物间的本质联系。数据库能有效地帮助一个组织或企业科学地管理各类信息资源。

数据是数据库中存储的基本对象，是按一定顺序排列组合的物理符号。数据有多种表现形式，可以是数字、文字、图像，甚至是音频或视频，它们都可以经过数字化后存入计算机。

数据库是数据的集合，具有统一的结构形式并存放于统一的存储介质内，是多种应用数据的集成，并可被各个应用程序所共享。

2. 数据的存储方式

计算机数据（Data）的存储一般以硬盘为数据资源存储空间，从而保证计算机内的数据能够持续保存。对于数据的处理，一般会采用数据库相关的技术进行处理，从而保证数据处理的高效性。

采用数据库的管理模式不仅提高了数据的存储效率，而且在存储的层面上提高了数据的安全性。通过分类的存储模式让数据管理更加安全便捷，更能实现对数据的调用和对比，并且方便查询等操作的使用。

3. 关系型数据库与非关系型数据库

（1）关系型数据库（Relational Database）。

关系型数据库是一种类型的数据库，其存储并提供访问被彼此相关的数据点。关系数据库基于关系模型，这是一种在表中表示数据的直观，直接的方法。在关系数据库中，表中的每一行都是一条记录，该记录具有唯一的 ID（称为 key）。该表的列保存数据的属性，每个记录通常为每个属性存储一个值，这使得在数据点之间建立关系变得容易。（例如：Oracle、DB2、PostgreSQL、Microsoft SQL Server、Microsoft Access、MySQL）。

（2）非关系型数据库（NoSQL）。

非关系型数据库指非关系型的，分布式的，且一般不保证遵循 ACID 原则的数据存储系统。非关系型数据库以键值对存储，且结构不固定，每一个元组可以有不一样的字段，每个元组可以根据需要增加一些自己的键值对，不局限于固定的结构，可以减少一些时间和空间的开销。（例如，列模型：Hbase；键值对模型：redis、MemcacheDB）。

 3.2 数据库服务运维

学习目标

学完本节后，您应能够：
- 掌握 MariaDB 服务初始化。
- 掌握 MariaDB 数据库运维。

3.2.1 初始化 MariaDB 服务

相较于 MySQL，MariaDB 数据库管理系统有了很多新鲜的扩展特性，例如，对微秒级别的支持、线程池、子查询优化、进程报告等。在配置妥当 yum 软件仓库后，即可安装部署 MariaDB 数据库主程序及服务端程序了。

（1）由于已经配置好 yum 源文件，此处只需直接使用命令安装即可。

```
[root@db ~]# yum -y install mariadb-server
```

（2）启动数据库服务，并设为开机自启。

```
[root@db ~]# systemctl start mariadb
[root@db ~]# systemctl enable mariadb
ln -s '/usr/lib/systemd/system/mariadb.service' '/etc/systemd/system/multi-user.target.wants/mariadb.service'
```

（3）在确认 MariaDB 数据库软件程序安装完毕并启动成功后不要立即使用。为了确保数据库的安全性和正常运转，需要先对数据库程序进行初始化操作。

```
[root@db ~]# mysql_secure_installation
/usr/bin/mysql_secure_installation: line 379: find_mysql_client: command not found
NOTE: RUNNING ALL PARTS OF THIS SCRIPT IS RECOMMENDED FOR ALL MariaDB
      SERVERS IN PRODUCTION USE!  PLEASE READ EACH STEP CAREFULLY!
In order to log into MariaDB to secure it, we'll need the current
password for the root user.  If you've just installed MariaDB, and
you haven't set the root password yet, the password will be blank,
so you should just press enter here.
Enter current password for root (enter for none):    #当前数据库密码为空，按回车键
OK, successfully used password, moving on...
Setting the root password ensures that nobody can log into the MariaDB
root user without the proper authorisation.
Set root password? [y/n] y                           #输入 y
New password:                                        #输入数据库 root 密码 000000
Re-enter new password:                               #再次输入密码 000000
Password updated successfully!
Reloading privilege tables..
 ... Success!
By default, a MariaDB installation has an anonymous user, allowing anyone
to log into MariaDB without having to have a user account created for
them.  This is intended only for testing, and to make the installation
go a bit smoother.  You should remove them before moving into a
production environment.
Remove anonymous users? [y/n] y                      #删除匿名账户
 ... Success!
Normally, root should only be allowed to connect from 'localhost'.  This
ensures that someone cannot guess at the root password from the network.
Disallow root login remotely? [y/n] n                #同意 root 管理员从远程登录
 ... skipping.
By default, MariaDB comes with a database named 'test' that anyone can
access.  This is also intended only for testing, and should be removed
before moving into a production environment.
Remove test database and access to it? [y/n] y       #删除 test 数据库并取消对它
```
的访问权限

```
 - Dropping test database...
 ... Success!
 - Removing privileges on test database...
 ... Success!
Reloading the privilege tables will ensure that all changes made so far
will take effect immediately.
Reload privilege tables now? [y/n] y      #刷新授权表,让初始化后的设定立即生效
 ... Success!
Cleaning up...
All done! If you've completed all of the above steps, your MariaDB
installation should now be secure.
Thanks for using MariaDB!
```

(4)一切准备就绪。现在我们将首次登录 MariaDB 数据库。其中,-u 参数用来指定以 root 管理员的身份登录,而 -p 参数用来验证该用户在数据库中的密码值。

```
[root@db ~]# mysql -uroot -p000000
Welcome to the MariaDB monitor.  Commands end with ; or \g.
Your MariaDB connection id is 11
Server version: 5.5.56-MariaDB MariaDB Server
Copyright (c) 2000, 2017, Oracle, MariaDB Corporation Ab and others.
Type 'help;' or '\h' for help. Type '\c' to clear the current input
statement.
MariaDB [(none)]>
```

(5)在登录 MariaDB 数据库后执行数据库命令时,都需要在命令后面用分号(;)结尾,这也是 Linux 命令最显著的区别。大家需要慢慢习惯数据库命令的这种设定。下面执行如下命令查看数据库管理系统中有哪些数据库:

```
MariaDB [(none)]> show databases;
+--------------------+
| Database           |
+--------------------+
| information_schema |
| mysql              |
| performance_schema |
+--------------------+
3 rows in set (0.00 sec)
```

3.2.2 MariaDB 数据库运维

1. 管理账户以及授权

在生产环境中,为了保障数据库系统的安全性,以及让其他用户协同管理数据库,我们可以在 MariaDB 数据库管理系统中为他们创建多个专用的数据库管理账户,然后再分配合理的权限,以满足他们的工作需求。

```
MariaDB [(none)]> create user data@localhost identified by '000000';
Query OK, 0 rows affected (0.00 sec)
```

(1)创建的账户信息可以使用 select 命令语句来查询。以下命令可查询 data 用户的主机名、账户名称以及加密过的密码值信息:

```
MariaDB [(none)]> use mysql
Reading table information for completion of table and column names
```

```
You can turn off this feature to get a quicker startup with -A
Database changed
MariaDB [mysql]> select host,user,password from user where user='data';
+-----------+------+-------------------------------------------+
| host      | user | password                                  |
+-----------+------+-------------------------------------------+
| localhost | data | *032197AE5731D4664921A6CCAC7CFCE6A0698693 |
+-----------+------+-------------------------------------------+
```

（2）数据库管理系统所使用的命令一般都比较复杂。我们以 grant 命令为例进行说明。grant 命令用于为账户进行授权，在使用 grant 命令时需要写上要赋予的权限、数据库及表单名称，以及对应的账户及主机信息。

赋予 data 用户访问本地数据库的所有权限。

```
MariaDB [(none)]> grant all privileges on *.* to data@'localhost' identified by '000000';
Query OK, 0 rows affected (0.00 sec)
```

然后使用 mysqladmin 命令更改 data 用户密码，并进行验证。

```
[root@db ~]# mysqladmin -udata -p000000 password '123456'
[root@db ~]# mysql -udata -p123456
Welcome to the MariaDB monitor.  Commands end with ; or \g.
Your MariaDB connection id is 28
Server version: 5.5.56-MariaDB MariaDB Server
Copyright (c) 2000, 2017, Oracle, MariaDB Corporation Ab and others.
Type 'help;' or '\h' for help. Type '\c' to clear the current input statement.
MariaDB [(none)]> show databases;
+--------------------+
| Database           |
+--------------------+
| information_schema |
| mysql              |
| performance_schema |
+--------------------+
3 rows in set (0.00 sec)
```

（3）指定授权用户，例如 data 用户有查询、增加、修改 mysql 数据库中所有表的权限，在执行完上述授权操作之后，我们再查看一下账户 data 的权限，操作命令如下：

```
[root@db ~]# mysql -uroot -p000000
Welcome to the MariaDB monitor.  Commands end with ; or \g.
Your MariaDB connection id is 30
Server version: 5.5.56-MariaDB MariaDB Server
Copyright (c) 2000, 2017, Oracle, MariaDB Corporation Ab and others.
Type 'help;' or '\h' for help. Type '\c' to clear the current input statement.
MariaDB [(none)]> grant select,update,insert on mysql.* to 'data'@'localhost' identified by '123456';
Query OK, 0 rows affected (0.00 sec)
MariaDB [(none)]> show grants for data@localhost;
+---------------------------------------------------------------------------------+
| Grants for data@localhost                                                       |
+---------------------------------------------------------------------------------+
| GRANT ALL PRIVILEGES ON *.* TO 'data'@'localhost' IDENTIFIED BY PASSWORD '*6BB4837EB74329105EE4568DDA7DC67ED2CA2AD9' |
```

```
| GRANT SELECT, INSERT, UPDATE ON 'mysql'.* TO 'data'@'localhost' |
+-----------------------------------------------------------------+
2 rows in set (0.00 sec)
```

（4）在 root 用户中移除刚才的授权。

```
MariaDB [(none)]> revoke select,update,insert on mysql.* from 'data'@'localhost';
Query OK, 0 rows affected (0.00 sec)
```

执行移除命令后，再查看账户 data 的信息。

```
MariaDB [(none)]> show grants for data@localhost;
+-----------------------------------------------------------------------+
| Grants for data@localhost                                             |
+-----------------------------------------------------------------------+
| GRANT USAGE ON *.* TO 'data'@'localhost' IDENTIFIED BY PASSWORD '*6BB483
7EB74329105EE4568DDA7DC67ED2CA2AD9' |
+-----------------------------------------------------------------------+
1 row in set (0.00 sec)
```

2. 数据库常用命令

MySQL 是一个简单的命令行 SQL 工具，该工具支持交互式和非交互式运行，使用 mysql 命令非常容易，下面就进入数据库常用命令的学习阶段。

（1）使用 show 命令查看数据库列表。

```
MariaDB [(none)]> show databases;
+--------------------+
| Database           |
+--------------------+
| information_schema |
| mysql              |
| performance_schema |
+--------------------+
3 rows in set (0.00 sec)
```

（2）切换使用 mysql 数据库。

```
MariaDB [(none)]> use mysql;
Reading table information for completion of table and column names
You can turn off this feature to get a quicker startup with -A

Database changed
```

（3）查看数据库中的表。

```
MariaDB [mysql]> show tables;
+---------------------------+
| Tables_in_mysql           |
+---------------------------+
| columns_priv              |
| db                        |
| event                     |
| func                      |
| general_log               |
| help_category             |
| help_keyword              |
| help_relation             |
| help_topic                |
```

```
| host                        |
| ndb_binlog_index            |
| plugin                      |
| proc                        |
| procs_priv                  |
| proxies_priv                |
| servers                     |
| slow_log                    |
| tables_priv                 |
| time_zone                   |
| time_zone_leap_second       |
| time_zone_name              |
| time_zone_transition        |
| time_zone_transition_type   |
| user                        |
+-----------------------------+
24 rows in set (0.00 sec)
```

（4）查看当前登录的用户。

```
MariaDB [mysql]> select user();
+----------------+
| user()         |
+----------------+
| root@localhost |
+----------------+
1 row in set (0.00 sec)
```

（5）查看当前数据库的版本。

```
MariaDB [mysql]> select version();
+----------------+
| version()      |
+----------------+
| 5.5.56-MariaDB |
+----------------+
1 row in set (0.00 sec)
```

（6）查看数据库状态并使用 like 和通配符。

```
MariaDB [mysql]> show status like'A%';
+----------------------------------+-------+
| Variable_name                    | Value |
+----------------------------------+-------+
| Aborted_clients                  | 0     |
| Aborted_connects                 | 6     |
| Access_denied_errors             | 0     |
| Aria_pagecache_blocks_not_flushed| 0     |
| Aria_pagecache_blocks_unused     | 15737 |
| Aria_pagecache_blocks_used       | 0     |
| Aria_pagecache_read_requests     | 0     |
| Aria_pagecache_reads             | 0     |
| Aria_pagecache_write_requests    | 0     |
| Aria_pagecache_writes            | 0     |
| Aria_transaction_log_syncs       | 0     |
+----------------------------------+-------+
11 rows in set (0.00 sec)
```

3. 创建管理数据库与表单

在 MariaDB 数据库管理系统中,一个数据库可以存放多个数据表,数据表单是数据库最重要、最核心的内容。用户可以根据需求自定义数据库表的结构,然后在其中合理地存放数据,以便后期轻松地维护和修改。

(1)建立数据库是管理数据的起点。现在尝试创建一个名为 data 的数据库,然后再查看数据库列表。

```
[root@db ~]# mysql -uroot -p000000
Welcome to the MariaDB monitor.  Commands end with ; or \g.
Your MariaDB connection id is 31
Server version: 5.5.56-MariaDB MariaDB Server
Copyright (c) 2000, 2017, Oracle, MariaDB Corporation Ab and others.
Type 'help;' or '\h' for help. Type '\c' to clear the current input statement.
MariaDB [(none)]> create database data;
Query OK, 1 row affected (0.00 sec)
MariaDB [(none)]> show databases;
+--------------------+
| Database           |
+--------------------+
| information_schema |
| data               |
| mysql              |
| performance_schema |
+--------------------+
4 rows in set (0.00 sec)
```

(2)创建数据库表单,首先要切换到指定的数据库中,然后定义相应的字段和长度以及数据信息。

```
MariaDB [(none)]> use data
Database changed
MariaDB [data]> create table user (name char(4),id int,sex int);
Query OK, 0 rows affected (0.05 sec)
MariaDB [data]> describe user;
+-------+---------+------+-----+---------+-------+
| Field | Type    | Null | Key | Default | Extra |
+-------+---------+------+-----+---------+-------+
| name  | char(4) | YES  |     | NULL    |       |
| id    | int(11) | YES  |     | NULL    |       |
| sex   | int(11) | YES  |     | NULL    |       |
+-------+---------+------+-----+---------+-------+
3 rows in set (0.00 sec)
```

(3)接下来向 user 数据表单中插入一条个人信息。此时需要用到 insert 命令,并在命令中写入相应的名称和对应的信息,最后使用 select 命令进行查询,需要加上想要查询的字段。

```
MariaDB [data]> insert into user(name,id,sex) values('jack','123','2');
Query OK, 1 row affected, 1 warning (0.02 sec)
MariaDB [data]> select * from user;
+------+------+------+
| name | id   | sex  |
+------+------+------+
| jack | 123  |    2 |
```

```
+------+------+------+
1 row in set (0.00 sec)
```

（4）在日常运维阶段，有时会出现用户的误报，如正确的 id 号为 321，此时需要修改刚刚创建的数据表单，我们使用 update 命令进行修改。

```
MariaDB [data]> update user set id=321;
Query OK, 1 row affected (0.03 sec)
Rows matched: 1  Changed: 1  Warnings: 0
MariaDB [data]> select * from user;
+------+------+------+
| name | id   | sex  |
+------+------+------+
| jack | 321  |   2  |
+------+------+------+
1 row in set (0.00 sec)
```

（5）还可以使用 delete 命令删除某个数据表单中的内容。下面使用 delete 命令删除数据表单 user 中的所有内容，然后再查看该表单的内容，可以发现该表单内容为空了。

```
MariaDB [data]> delete from user;
Query OK, 1 row affected (0.03 sec)
MariaDB [data]> select * from user;
Empty set (0.00 sec)
```

4. 数据库的备份与恢复

mysqldump 命令用于备份数据库数据。其中参数与 mysql 命令大致相同，-u 参数代表用户定义登录数据库的账户名称，-p 参数代表密码提示符。

（1）下面将 data 数据库中的内容导出成一个文件，并保存到 root 管理员的家目录下面（使用上一步的实验先写入一个数据进行测试）。

```
[root@db ~]# mysqldump -u root -p  data > /root/data.dump
Enter password:
[root@db ~]# ls
data.dump
```

（2）进入 MariaDB 数据库管理系统，彻底删除 data 数据库并且数据表也会随之删除，然后创建一个名为 data 的新数据库，此时里面没有数据。

```
MariaDB [(none)]> drop database data;
Query OK, 1 row affected (0.02 sec)
MariaDB [(none)]> create database data;
Query OK, 1 row affected (0.00 sec)
```

使用输入重定向把刚刚备份的数据库文件导入数据库系统中，然后登录 MariaDB 数据库，就可以查看到之前备份的数据表单，说明数据库成功恢复。

```
[root@db ~]# mysql -uroot -p data < /root/data.dump
Enter password:
[root@db ~]# mysql -uroot -p000000
Welcome to the MariaDB monitor.  Commands end with ; or \g.
Your MariaDB connection id is 35
Server version: 5.5.56-MariaDB MariaDB Server
Copyright (c) 2000, 2017, Oracle, MariaDB Corporation Ab and others.
Type 'help;' or '\h' for help. Type '\c' to clear the current input statement.
MariaDB [(none)]> use data;
```

```
Reading table information for completion of table and column names
You can turn off this feature to get a quicker startup with -A
Database changed
MariaDB [data]> show tables;
+-----------------+
| Tables_in_data  |
+-----------------+
| user            |
+-----------------+
1 row in set (0.00 sec)
MariaDB [data]> select * from user;
+------+------+------+
| name | id   | sex  |
+------+------+------+
| Tom  | 999  |   1  |
+------+------+------+
1 row in set (0.00 sec)
```

3.3 主从复制数据库

学习目标

学完本节后，您应能够：
- 了解主从复制数据库服务。
- 掌握主从复制数据库部署。

3.3.1 主从复制数据库服务

1. MySQL 主从架构简介

MySQL 支持几种不同形式的复制，如单双向、链式级联、异步复制。在复制过程中，一个服务器充当主服务器（master），而一个或多个其他的服务器充当从服务器（slave）如果设置了链式级联复制，那么从 slave 服务器除了充当从服务器，还会作为其他服务器的主服务器。

当配好主从复制后，数据库的内容更新应该都在 master 服务器上，避免用户对主服务器上数据库内容的更新与对从服务器上数据库内容的更新产生冲突。生产环境中一般会忽略授权表的同步，对在从服务器上的用户仅授权 select 读权限。或在 my.cnf 配置文件中加 read-only 参数，来确保从数据库只读，当然二者同时操作效果更佳。

2. 主从复制的意义

使用 MySQL 复制功能可以将主服务器上的数据复制到多台从服务器上。默认情况下，复制是异步传输方式，从服务器不需要总是连接主服务器去更新数据。也就是说，数据更新可以在远距离连接的情况下进行，甚至在使用拨号网络的临时连接环境下也可以进行。根据自定义设置，我们可以对所有的数据库或部分数据库甚至是部分数据表进行复制。通过主从复制在企业级应用环境中就不必再担心数据库的单点故障，当一台服务器宕机时，其他服务器一样可以提供非常稳定、可靠的数据服务。

3. 主从复制的优点

MySQL 的主从复制是用来建立一个和主数据库完全一样的数据库环境，称为从数据库，主数据库一般是实时的业务数据操作，从数据库常以读取为主。优点主要包括：

（1）可以作为备用数据库进行操作，当主数据库出现故障之后，从数据库可以替代主数据库继续工作，不影响业务流程。

（2）读写分离，将读和写应用在不同的数据库与服务器上。一般读写的数据库环境配置为，一个写入的数据库，一个或多个读的数据库，各个数据库分别位于不同的服务器上，充分利用服务器性能和数据库性能；当然，其中会涉及如何保证读写数据库的数据一致，这样就可以利用主从复制技术来完成。

（3）吞吐量较大，业务的查询较多，并发与负载较大。

3.3.2　任务：部署主从复制数据库

通过主从复制在企业级应用环境中就不必再担心数据库的单点故障，当一台服务器宕机时，其他服务器一样可以提供非常稳定、可靠的数据服务。

任务执行清单

在本任务中，您将掌握 MySQL 主从复制服务的架构部署。

目标

- 掌握 MySQL 主从服务的部署。

重要信息

- 在此实验之前已经安装好数据库服务器。
- 所有的云主机密码都为 000000。

解决方案

1. 配置主数据库

（1）修改 mysql1 节点的数据库配置文件，在配置文件 /etc/my.cnf 中的 [mysqld] 增添如下三行内容。

```
[root@mysql1 ~]# cat /etc/my.cnf
[mysqld]
log_bin = mysql-bin                    # 记录操作日志
binlog_ignore_db = mysql               # 不同步 mysql 系统数据库
server_id = 14                         # 数据库集群中的每个节点 id 都要不同，
一般使用 IP 地址的最后段的数字，例如 192.168.200.14，server_id 就写 14
```

（2）重启数据库服务，并进入数据库。

```
[root@mysql1 ~]# systemctl restart mariadb
[root@mysql1 ~]# mysql -uroot -p000000
Welcome to the MariaDB monitor.  Commands end with ; or \g.
Your MariaDB connection id is 2
Server version: 5.5.56-MariaDB MariaDB Server
Copyright (c) 2000, 2017, Oracle, MariaDB Corporation Ab and others.
Type 'help;' or '\h' for help. Type '\c' to clear the current input statement.
MariaDB [(none)]>
```

(3)在 mysql1 节点，授权在任何客户端机器上可以以 root 用户登录到数据库，然后在主节点上创建一个 user 用户连接节点 mysql2，并赋予从节点同步主节点数据库的权限。

```
MariaDB [(none)]> grant all privileges on *.* to root@'%' identified by "000000";
Query OK, 0 rows affected (0.00 sec)
MariaDB [(none)]> grant replication slave on *.* to 'user'@'mysql2' identified by '000000';
Query OK, 0 rows affected (0.00 sec)
```

2. 配置从数据库

（1）修改 mysql2 节点的数据库配置文件，在配置文件 /etc/my.cnf 中的 [mysqld] 增添如下三行内容。

```
[root@mysql2 ~]# cat /etc/my.cnf
[mysqld]
log_bin = mysql-bin          # 记录操作日志
binlog_ignore_db = mysql     # 不同步 mysql 系统数据库
server_id = 15               # 数据库集群中的每个节点 id 都要不同，一般使用 IP 地址的最后段的数字，例如 192.168.200.15，server_id 就写 15
```

（2）在从节点 mysql2 上登录 MariaDB 数据库，配置从节点连接主节点的连接信息。master_host 为主节点主机名 mysql1，master_user 为上一步中创建的用户 user。

```
[root@mysql2 ~]# systemctl restart mariadb
[root@mysql2 ~]# mysql -uroot -p000000
Welcome to the MariaDB monitor.  Commands end with ; or \g.
Your MariaDB connection id is 2
Server version: 5.5.56-MariaDB MariaDB Server
Copyright (c) 2000, 2017, Oracle, MariaDB Corporation Ab and others.
Type 'help;' or '\h' for help. Type '\c' to clear the current input statement.
MariaDB [(none)]> change master to master_host='mysql1',master_user='user',master_password='000000';
Query OK, 0 rows affected (0.01 sec)
```

（3）配置完毕主从数据库之间的连接信息之后，开启从节点服务。使用 show slave status\G 命令，并查看从节点服务状态，如果 Slave_IO_Running 和 Slave_SQL_Running 的状态都为 Yes，则从节点服务开启成功。

```
MariaDB [(none)]> start slave;
MariaDB [(none)]> show slave status\G;
*************************** 1. row ***************************
               Slave_IO_State: Waiting for master to send event
                  Master_Host: mysql1
                  Master_User: user
                  Master_Port: 3306
                Connect_Retry: 60
              Master_Log_File: mysql-bin.000001
          Read_Master_Log_Pos: 531
               Relay_Log_File: mariadb-relay-bin.000002
                Relay_Log_Pos: 815
        Relay_Master_Log_File: mysql-bin.000001
             Slave_IO_Running: Yes
            Slave_SQL_Running: Yes
```

```
                   Replicate_Do_DB:
               Replicate_Ignore_DB:
                Replicate_Do_Table:
            Replicate_Ignore_Table:
           Replicate_Wild_Do_Table:
       Replicate_Wild_Ignore_Table:
                        Last_Errno: 0
                        Last_Error:
                      Skip_Counter: 0
               Exec_Master_Log_Pos: 531
                   Relay_Log_Space: 1111
                   Until_Condition: None
                    Until_Log_File:
                     Until_Log_Pos: 0
                Master_SSL_Allowed: No
                Master_SSL_CA_File:
                Master_SSL_CA_Path:
                   Master_SSL_Cert:
                 Master_SSL_Cipher:
                    Master_SSL_Key:
             Seconds_Behind_Master: 0
     Master_SSL_Verify_Server_Cert: No
                     Last_IO_Errno: 0
                     Last_IO_Error:
                    Last_SQL_Errno: 0
                    Last_SQL_Error:
         Replicate_Ignore_Server_Ids:
                  Master_Server_Id: 14
```

可以看到 Slave_IO_Running 和 Slave_SQL_Running 的状态都是 Yes，配置数据库主从集群成功。

本任务到此结束。

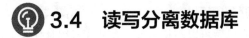 3.4 读写分离数据库

学习目标

学完本节后，您应能够：

- 了解数据库读写分离服务。
- 了解 Mycat 中间件服务。

3.4.1 读写分离数据库服务

1. 读写分离简介

读写分离，基本的原理是让主数据库处理事务性增、改、删操作（insert、update、delete），而从数据库处理 select 查询操作。数据库复制被用来把事务性操作导致的变更同步到集群中的从数据库。

2. 读写分离的作用

因为数据库的"写"（写 10 000 条数据到 MySQL 可能要 3 min）操作是比较耗时的，但是数据库的"读"（从 MySQL 读 10 000 条数据可能只要 5 s）。所以读写分离可以解决数据库写入时影响查询效率的问题。

3. 读写分离应用场景

数据库不一定总要读写分离，如果程序使用数据库较多、更新较少、查询较多的情况下会考虑使用。利用数据库主从同步，可以减少数据库压力，提高性能。当然，数据库也有其他优化方案。例如使用 memcache、分表、搜索引擎等方法。

4. 主从复制、读写分离的基本设计

在实际的生产环境中，对数据库的读和写都在同一个数据库服务器中，是不能满足实际需求的。无论是在安全性、高可用性还是高并发等各个方面都是完全不能满足实际需求的。因此，通过主从复制的方式来同步数据，再通过一台主、多台从节点，主节点提供写操作，从节点提供读操作来实现读写分离，从而提升数据库的并发负载能力。

3.4.2 Mycat 中间件服务

1. 读写分离的好处

（1）分摊服务器压力，提高机器的系统处理效率。

读写分离适用于读远比写多的场景，如果有一台服务器，当 select 很多时，update 和 delete 会被这些 select 访问中的数据堵塞，等待 select 结束，并发性能并不高，而主从只负责各自的写和读，极大程度地缓解了 X 锁和 S 锁争用。

假如我们有 1 主 3 从，不考虑上述 1 中提到的从库单方面设置，假设现在 1 min 内有 10 条写入，150 条读取。那么，1 主 3 从相当于共计 40 条写入，而读取总数没变，因此平均下来每台服务器承担了 10 条写入和 50 条读取（主库不承担读取操作）。因此，虽然写入没变，但是读取大大分摊了，提高了系统性能。另外，当读取被分摊后，又间接提高了写入的性能。所以，总体性能提高了，简言之就是用机器和带宽换性能。

（2）增加冗余，提高服务可用性。

当一台数据库服务器宕机后可以调整另外一台从库，以最快速度恢复服务。

2. Mycat 简介

Mycat 是一个开源的分布式数据库系统，但是因为数据库一般都有自己的数据库引擎，而 Mycat 并没有属于自己的独有数据库引擎，所有严格意义上说并不能算是一个完整的数据库系统，只能说是一个在应用和数据库之间起桥梁作用的中间件。

在 Mycat 中间件出现之前，MySQL 主从复制集群，如果要实现读写分离，一般是在程序段实现，这样就带来了一个问题，即数据段和程序的耦合度太高，如果数据库的地址发生了改变，那么程序也要进行相应的修改，如果数据库出故障了，则同时也意味着程序的不可用，而对于很多应用来说，并不能接受。

3. Mycat 中间件优势

引入 Mycat 中间件能很好地对程序和数据库进行解耦，这样，程序只需关注数据库中间件的地址，而无须知晓底层数据库是如何提供服务的，大量的通用数据聚合、事务、数据源切换等工作都由中间件来处理。

Mycat 中间件的原理是对数据进行分片处理，从原有的一个库，被切分为多个分片数据库，所有的分片数据库集群构成完成的数据库存储，有点类似磁盘阵列中的 RAID0.

3.4.3 任务：配置 Mycat 数据库中间件

MyCat 就是实现数据库集群的，对海量数据的数据存储的一种解决方案，因为很多数据库不想 Oracle 一样自带集群的配置，那么在进行海量数据存储的时候就要使用到 MyCat 进行数据库的管理了。

任务执行清单

在本任务中，您将掌握 Mycat 数据库中间件的配置。

目标

- 掌握 Mycat 服务的安装与配置。

重要信息

- 新建一台新的虚拟机。
- 云主机密码为 000000。

解决方案

1. 登录虚拟机，安装 Mycat 服务

（1）安装 Mycat 服务，将 Mycat 服务的二进制软件包 Mycat-server-1.6-RELEASE-20161028204710-linux.tar.gz 上传到 Mycat 虚拟机的 /root 目录下，并将软件包解压到 /use/local 目录中。赋予解压后的 Mycat 目录权限。最后在 /etc/profile 系统变量文件中添加 Mycat 服务的系统变量，并生效变量。

```
[root@mycat ~]# tar -zxvf Mycat-server-1.6-RELEASE-20161028204710-linux.tar.gz  -C /usr/local/
[root@mycat ~]# chown -R 777 /usr/local/mycat/
[root@mycat ~]# echo export MYCAT_HOME=/usr/local/mycat/ >>  /etc/profile
[root@mycat ~]# source /etc/profile
```

（2）部署 Mycat 中间件服务需要先部署 JDK 1.7 或以上版本的 JDK 软件环境，这里部署 JDK 1.8 版本。

Mycat 节点安装 Java 环境：

```
[root@mycat ~]# yum install -y java-1.8.0-openjdk java-1.8.0-openjdk-devel
[root@mycat ~]# java -version
openjdk version "1.8.0_161"
OpenJDK Runtime Environment (build 1.8.0_161-b14)
OpenJDK 64-Bit Server VM (build 25.161-b14, mixed mode)
```

2. 编辑 Mycat 的逻辑库配置文件

（1）配置 Mycat 服务读写分离的 schema.xml 配置文件在 /usr/local/mycat/conf/ 目录下，可以在文件中定义一个逻辑库，使用户可以通过 Mycat 服务管理该逻辑库对应的 MariaDB 数据库。在这里定义一个逻辑库 schema，name 为 USERDB；该逻辑库 USERDB 对应数据库 database 为 test（在部署主从数据库时已安装）；设置数据库写入节点为主节点 mysql1；设置数据库读取节点为从节点 mysql2。可以直接删除原来 schema.xml 的内容，替换为如下内容：

> **注意**：IP 需要修改成实际的 IP 地址。

```
[root@mycat ~]# cat /usr/local/mycat/conf/schema.xml
<?xml version="1.0"?>
<!DOCTYPE mycat:schema SYSTEM "schema.dtd">
<mycat:schema xmlns:mycat="http://io.mycat/">
    <schema name="USERDB" checkSQLschema="true" sqlMaxLimit="100" dataNode="dn1"></schema>
    <dataNode name="dn1" dataHost="localhost1" database="test" />
    <dataHost name="localhost1" maxCon="1000" minCon="10" balance="3" dbType="mysql" dbDriver="native" writeType="0" switchType="1" slaveThreshold="100">
        <heartbeat>select user()</heartbeat>
        <writeHost host="hostM1" url="192.168.200.14:3306" user="root" password="000000">
            <readHost host="hostS1" url="192.168.200.15:3306" user="root" password="000000" />
        </writeHost>
    </dataHost>
</mycat:schema>
```

（2）修改配置文件权限，然后编辑 Mycat 的访问用户，在 /usr/local/mycat/conf/server.xml 中，修改 root 用户的访问密码与数据库的密码为 000000，访问 Mycat 的逻辑库为 USERDB。

```
[root@mycat ~]# chown root:root /usr/local/mycat/conf/schema.xml
[root@mycat ~]# vi /usr/local/mycat/conf/server.xml
# 在配置文件的最后部分：
<user name="root">
        <property name="password">000000</property>
        <property name="schemas">USERDB</property>
# 删除如下几行：
<user name="user">
        <property name="password">user</property>
        <property name="schemas">TESTDB</property>
        <property name="readOnly">true</property>
</user>
```

（3）启动 Mycat 服务，通过命令启动 Mycat 数据库中间件服务，启动后使用 netstat -ntpl 命令查看虚拟机端口开放情况，如果有开放 8066 和 9066 端口，则表示 Mycat 服务开启成功，如图 3-2 所示。

```
[root@mycat ~]# /bin/bash /usr/local/mycat/bin/mycat start
```

```
[root@mycat ~]# netstat -ntpl
Active Internet connections (only servers)
Proto Recv-Q Send-Q Local Address           Foreign Address         State       PID/Program name
tcp        0      0 0.0.0.0:22              0.0.0.0:*               LISTEN      943/sshd
tcp        0      0 127.0.0.1:25            0.0.0.0:*               LISTEN      862/master
tcp        0      0 127.0.0.1:32000         0.0.0.0:*               LISTEN      1414/java
tcp6       0      0 :::8066                 :::*                    LISTEN      1414/java
tcp6       0      0 :::9066                 :::*                    LISTEN      1414/java
tcp6       0      0 :::35155                :::*                    LISTEN      1414/java
tcp6       0      0 :::22                   :::*                    LISTEN      943/sshd
tcp6       0      0 :::34744                :::*                    LISTEN      1414/java
tcp6       0      0 ::1:25                  :::*                    LISTEN      862/master
tcp6       0      0 :::1984                 :::*                    LISTEN      1414/java
```

图 3-2 查询 Mycat 服务端口

由此，Mycat 服务配置完成，本任务到此结束。

3.5 开放研究任务:构建读写分离数据库集群

使用一台虚拟机部署 Mycat 数据库中间件服务,将用户提交的读写操作识别分发给相应的数据库节点。这样将用户的访问操作、数据库的读与写操作分给 3 台主机,只有数据库集群的主节点接收增、删、改 SQL 语句,从节点接收查询语句,分担了主节点的查询压力。

任务执行清单

在本任务中,您将通过实践使用 Mycat 作为数据库中间件服务构建读写分离的数据库集群。
(1)配置基础环境。
(2)主从服务器部署。
(3)Mycat 服务部署。
(4)数据库集群构建。

目标

- 掌握主从数据库的部署。
- 掌握 Mycat 中间件配置。
- 掌握读写数据库集群部署。

重要信息

- 虚拟机为 CentOS 7 的操作系统。
- 虚拟机镜像使用 CentOS 7.1804 版本。
- 虚拟机密码为 000000。

解决方案

1. 配置基础环境

(1)使用远程连接工具 CRT 连接到 192.168.200.14、192.168.200.15、192.168.200.16 这三台云主机,并对三台虚拟机进行修改主机名的操作,分别为 mysql1、myslq2、mycat。此处以 mysql1 节点为例。

mysql1 节点:

```
[root@localhost ~]# hostnamectl set-hostname mysql1
[root@localhost ~]# logout
[root@mysql1 ~]# hostnamectl
   Static hostname: mysql1
         Icon name: computer-vm
           Chassis: vm
        Machine ID: 179f6c8f2e7942ef81b0f5565a6883fa
           Boot ID: 69ad020d53e54892b9005f82e182c140
    Virtualization: kvm
  Operating System: CentOS Linux 7 (Core)
       CPE OS Name: cpe:/o:centos:centos:7
            Kernel: Linux 3.10.0-327.el7.x86_64
      Architecture: x86-64
```

(2)配置 hosts 文件。在三个节点配置 /etc/hosts 文件,修改为如下(此处以 mysql1 节点为例):

```
[root@mysql1 ~]# cat /etc/hosts
127.0.0.1       localhost localhost.localdomain localhost4 localhost4.localdomain4
::1             localhost localhost.localdomain localhost6 localhost6.localdomain6
192.168.200.14  mysql1
192.168.200.15  mysql2
```

（3）安装数据库。在配置妥当 yum 软件仓库后，即可安装部署 MariaDB 数据库主程序及服务端程序了，在两个节点安装数据库服务，此处以 mysql1 节点为例：

```
[root@mysql1 ~]# yum install -y mariadb mariadb-server
```

两个节点启动数据库服务并设置开机自启，此处以 mysql1 节点为例：

```
[root@mysql1 ~] # systemctl start mariadb
[root@mysql1 ~] # systemctl enable mariadb
Created symlink from /etc/systemd/system/multi-user.target.wants/mariadb.service to /usr/lib/systemd/system/mariadb.service.
```

（4）初始化 MariaDB 服务。两个节点初始化数据库，配置数据库 root 密码为 000000，此处以 mysql1 节点为例：

```
[root@mysql1 ~]# mysql_secure_installation
/usr/bin/mysql_secure_installation: line 379: find_mysql_client: command not found
NOTE: RUNNING ALL PARTS OF THIS SCRIPT IS RECOMMENDED FOR ALL MariaDB
      SERVERS IN PRODUCTION USE!  PLEASE READ EACH STEP CAREFULLY!
In order to log into MariaDB to secure it, we'll need the current
password for the root user.  If you've just installed MariaDB, and
you haven't set the root password yet, the password will be blank,
so you should just press enter here.
Enter current password for root (enter for none):      #默认按回车键
OK, successfully used password, moving on...
Setting the root password ensures that nobody can log into the MariaDB
root user without the proper authorisation.
Set root password? [Y/n] y
New password:                                          #输入数据库 root 密码 000000
Re-enter new password:                                 #再次输入密码 000000
Password updated successfully!
Reloading privilege tables..
 ... Success!
By default, a MariaDB installation has an anonymous user, allowing anyone
to log into MariaDB without having to have a user account created for
them.  This is intended only for testing, and to make the installation
go a bit smoother.  You should remove them before moving into a
production environment.
Remove anonymous users? [Y/n] y
 ... Success!
Normally, root should only be allowed to connect from 'localhost'.  This
ensures that someone cannot guess at the root password from the network.
Disallow root login remotely? [Y/n] n
 ... skipping.
By default, MariaDB comes with a database named 'test' that anyone can
access.  This is also intended only for testing, and should be removed
before moving into a production environment.
Remove test database and access to it? [Y/n] y
 - Dropping test database...
 ... Success!
```

```
  - Removing privileges on test database...
   ... Success!
 Reloading the privilege tables will ensure that all changes made so far
 will take effect immediately.
 Reload privilege tables now? [Y/n] y
  ... Success!
 Cleaning up...
 All done!  If you've completed all of the above steps, your MariaDB
 installation should now be secure.
 Thanks for using MariaDB!
```

2. 主从服务器部署

(1) 修改 mysql1 节点的数据库配置文件，在配置文件 /etc/my.cnf 中的 [mysqld] 增添如下三行内容。

```
[root@mysql1 ~]# cat /etc/my.cnf
[mysqld]
log_bin=mysql-bin                  #记录操作日志
binlog_ignore_db=mysql             #不同步mysql系统数据库
server_id=14                       #数据库集群中的每个节点id都要不同，一
#般使用IP地址的最后段的数字，例如192.168.200.14，server_id就写14
```

(2) 重启数据库服务，并进入数据库。

```
[root@mysql1 ~]# systemctl restart mariadb
[root@mysql1 ~]# mysql -uroot -p000000
Welcome to the MariaDB monitor.  Commands end with ; or \g.
Your MariaDB connection id is 2
Server version: 5.5.56-MariaDB MariaDB Server
Copyright (c) 2000, 2017, Oracle, MariaDB Corporation Ab and others.
Type 'help;' or '\h' for help. Type '\c' to clear the current input
statement.
MariaDB [(none)]>
```

(3) 在 mysql1 节点，授权在任何客户端机器上可以以 root 用户登录到数据库，然后在主节点上创建一个 user 用户连接节点 mysql2，并赋予从节点同步主节点数据库的权限。

```
MariaDB [(none)]> grant all privileges  on *.* to root@'%' identified by
"000000";
Query OK, 0 rows affected (0.00 sec)
MariaDB [(none)]> grant replication slave on *.* to 'user'@'mysql2'
identified by '000000';
Query OK, 0 rows affected (0.00 sec)
```

(4) 修改 mysql2 节点的数据库配置文件，在配置文件 /etc/my.cnf 中的 [mysqld] 增添如下三行内容。

```
[root@mysql2 ~]# cat /etc/my.cnf
[mysqld]
log_bin=mysql-bin                  #记录操作日志
binlog_ignore_db=mysql             #不同步mysql系统数据库
server_id=15                       #数据库集群中的每个节点id都要不同，一般使用
#IP地址的最后段的数字，例如192.168.200.15，server_id就写15
```

(5) 在从节点 mysql2 上登录 MariaDB 数据库，配置从节点连接主节点的连接信息。master_host 为主节点主机名 mysql1，master_user 为上一步中创建的用户 user。

```
[root@mysql2 ~]# systemctl restart mariadb
[root@mysql2 ~]# mysql -uroot -p000000
Welcome to the MariaDB monitor.  Commands end with ; or \g.
Your MariaDB connection id is 2
Server version: 5.5.56-MariaDB MariaDB Server
Copyright (c) 2000, 2017, Oracle, MariaDB Corporation Ab and others.
Type 'help;' or '\h' for help. Type '\c' to clear the current input statement.
MariaDB [(none)]> change master to master_host='mysql1',master_user='user',
master_password='000000';
Query OK, 0 rows affected (0.01 sec)
```

（6）配置完毕主从数据库之间的连接信息之后，开启从节点服务。使用 show slave status\G 命令，并查看从节点服务状态，如果 Slave_IO_Running 和 Slave_SQL_Running 的状态都为 YES，则从节点服务开启成功。

```
MariaDB [(none)]> start slave;
MariaDB [(none)]> show slave status\G;
*************************** 1. row ***************************
               Slave_IO_State: Waiting for master to send event
                  Master_Host: mysql1
                  Master_User: user
                  Master_Port: 3306
                Connect_Retry: 60
              Master_Log_File: mysql-bin.000001
          Read_Master_Log_Pos: 531
               Relay_Log_File: mariadb-relay-bin.000002
                Relay_Log_Pos: 815
        Relay_Master_Log_File: mysql-bin.000001
             Slave_IO_Running: Yes
            Slave_SQL_Running: Yes
              Replicate_Do_DB:
          Replicate_Ignore_DB:
           Replicate_Do_Table:
       Replicate_Ignore_Table:
      Replicate_Wild_Do_Table:
  Replicate_Wild_Ignore_Table:
                   Last_Errno: 0
                   Last_Error:
                 Skip_Counter: 0
          Exec_Master_Log_Pos: 531
              Relay_Log_Space: 1111
              Until_Condition: None
               Until_Log_File:
                Until_Log_Pos: 0
           Master_SSL_Allowed: No
           Master_SSL_CA_File:
           Master_SSL_CA_Path:
              Master_SSL_Cert:
            Master_SSL_Cipher:
               Master_SSL_Key:
        Seconds_Behind_Master: 0
Master_SSL_Verify_Server_Cert: No
                Last_IO_Errno: 0
```

```
            Last_IO_Error:
            Last_SQL_Errno: 0
            Last_SQL_Error:
  Replicate_Ignore_Server_Ids:
            Master_Server_Id: 14
```

（7）验证数据库主从服务。先在主节点 mysql1 中创建库 test，并在库 test 中创建表 company，插入表数据，创建完成后，查看表 company 数据，命令如下所示：

```
[root@mysql1 ~]# mysql -uroot -p000000
Welcome to the MariaDB monitor.  Commands end with ; or \g.
Your MariaDB connection id is 2
Server version: 5.5.56-MariaDB MariaDB Server
Copyright (c) 2000, 2017, Oracle, MariaDB Corporation Ab and others.
Type 'help;' or '\h' for help. Type '\c' to clear the current input statement.
MariaDB [(none)]> create database test;
Query OK, 1 row affected (0.00 sec)
MariaDB [(none)]> use test;
Database changed
MariaDB [test]> create table company(id int not null primary key,name varchar(50),addr varchar(255));
Query OK, 0 rows affected (0.01 sec)
MariaDB [test]> insert into company values(1,"alibaba","china");
Query OK, 1 row affected (0.01 sec)
MariaDB [test]> select * from company;
+----+---------+-------+
| id | name    | addr  |
+----+---------+-------+
|  1 | alibaba | china |
+----+---------+-------+
1 row in set (0.00 sec)
```

（8）登录 mysql2 节点的数据库，查看数据库列表。找到 test 数据库，查询表，并查询内容，验证从数据库的复制功能，命令如下所示：

```
[root@mysql2 ~]# mysql -uroot -p000000
Welcome to the MariaDB monitor.  Commands end with ; or \g.
Your MariaDB connection id is 2
Server version: 5.5.56-MariaDB MariaDB Server
Copyright (c) 2000, 2017, Oracle, MariaDB Corporation Ab and others.
Type 'help;' or '\h' for help. Type '\c' to clear the current input statement.
MariaDB [(none)]> show databases;
+--------------------+
| Database           |
+--------------------+
| information_schema |
| mysql              |
| performance_schema |
| test               |
+--------------------+
4 rows in set (0.00 sec)
MariaDB [(none)]> use test;
Reading table information for completion of table and column names
```

```
You can turn off this feature to get a quicker startup with -A
Database changed
MariaDB [test]> show tables;
+----------------+
| Tables_in_test |
+----------------+
| company        |
+----------------+
1 row in set (0.00 sec)
MariaDB [test]> select * from company;
+----+---------+-------+
| id | name    | addr  |
+----+---------+-------+
|  1 | alibaba | china |
+----+---------+-------+
1 row in set (0.00 sec)
```

可以查看到主数据库中刚刚创建的库、表、信息，验证从数据库的复制功能成功。

3. Mycat 服务部署

（1）安装 Mycat 服务，赋予解压后的 Mycat 目录权限。最后在 /etc/profile 系统变量文件中添加 Mycat 服务的系统变量，并生效变量。

```
[root@mycat ~]# tar -zxvf Mycat-server-1.6-RELEASE-20161028204710-linux.tar.gz  -C /usr/local/
[root@mycat ~]# chown -R 777 /usr/local/mycat/
[root@mycat ~]# echo export MYCAT_HOME=/usr/local/mycat/ >>  /etc/profile
[root@mycat ~]# source /etc/profile
```

（2）部署 Mycat 中间件服务需要先部署 JDK 1.7 或以上版本的 JDK 软件环境，这里部署 JDK 1.8 版本。

Mycat 节点安装 Java 环境：

```
[root@mycat ~]# yum install -y java-1.8.0-openjdk java-1.8.0-openjdk-devel
[root@mycat ~]# java -version
openjdk version "1.8.0_161"
OpenJDK Runtime Environment (build 1.8.0_161-b14)
OpenJDK 64-Bit Server VM (build 25.161-b14, mixed mode)
```

（3）编辑 Mycat 的逻辑库 schema.xml 配置文件。可以直接删除原来 schema.xml 的内容，替换为如下：

```
[root@mycat ~]# cat /usr/local/mycat/conf/schema.xml
<?xml version="1.0"?>
<!DOCTYPE mycat:schema SYSTEM "schema.dtd">
<mycat:schema xmlns:mycat="http://io.mycat/">
    <schema name="USERDB" checkSQLschema="true" sqlMaxLimit="100" dataNode="dn1"></schema>
    <dataNode name="dn1" dataHost="localhost1" database="test" />
    <dataHost name="localhost1" maxCon="1000" minCon="10" balance="3" dbType="mysql" dbDriver="native" writeType="0" switchType="1"  slaveThreshold="100">
        <heartbeat>select user()</heartbeat>
        <writeHost host="hostM1" url="192.168.200.14:3306" user="root" password="000000">
```

```
            <readHost host="hostS1" url="192.168.200.15:3306" user="root"
password="000000" />
        </writeHost>
    </dataHost>
</mycat:schema>
```

注意：IP 需要修改成实际的 IP 地址。

（4）修改配置文件权限，然后编辑 Mycat 的访问用户，配置如下所示：

```
[root@mycat ~]# chown root:root /usr/local/mycat/conf/schema.xml
[root@mycat ~]# vi /usr/local/mycat/conf/server.xml
# 在配置文件的最后部分:
<user name="root">
        <property name="password">000000</property>
        <property name="schemas">USERDB</property>
# 删除如下几行:
<user name="user">
        <property name="password">user</property>
        <property name="schemas">TESTDB</property>
        <property name="readOnly">true</property>
</user>
```

（5）通过命令启动 Mycat 数据库中间件服务，并使用 netstat -ntpl 命令查看虚拟机是否开放 8066 和 9066 端口。

```
[root@mycat ~]# /bin/bash /usr/local/mycat/bin/mycat start
```

（6）验证数据库集群服务读写分离功能。在 Mycat 虚拟机中安装 MariaDB-client 服务，使用 mysql 命令查看 Mycat 服务的逻辑库 USERDB。

```
[root@mycat ~]# yum install MariaDB-client
[root@mycat ~]# mysql -h 127.0.0.1 -P 8066 -uroot -p000000
Welcome to the MariaDB monitor.  Commands end with ; or \g.
Your MySQL connection id is 1
Server version: 5.6.29-mycat-1.6-RELEASE-20161028204710 MyCat Server
(OpenCloundDB)
Copyright (c) 2000, 2018, Oracle, MariaDB Corporation Ab and others.
Type 'help;' or '\h' for help. Type '\c' to clear the current input
statement.
MySQL [(none)]> show databases;
+----------+
| DATABASE |
+----------+
| USERDB   |
+----------+
1 row in set (0.004 sec)
MySQL [(none)]> use USERDB
Reading table information for completion of table and column names
You can turn off this feature to get a quicker startup with -A
Database changed
MySQL [USERDB]> show tables;
+----------------+
| Tables_in_test |
+----------------+
| company        |
```

```
+----------------+
1 row in set (0.002 sec)
MySQL [USERDB]> select * from company;
+----+---------+-------+
| id | name    | addr  |
+----+---------+-------+
|  1 | alibaba | china |
+----+---------+-------+
1 row in set (0.003 sec)
```

（7）用 Mycat 服务添加数据表，在 Mycat 虚拟机上使用 mysql 命令对表 company 添加一条数据 (2,'TX', 'shenzhen')，添加完毕后查看表信息。命令如下所示：

```
MySQL [USERDB]> insert into company values (2,'TX','shenzhen');
Query OK, 1 row affected (0.031 sec)
MySQL [USERDB]> select * from company;
+----+---------+----------+
| id | name    | addr     |
+----+---------+----------+
|  1 | alibaba | china    |
|  2 | TX      | shenzhen |
+----+---------+----------+
2 rows in set (0.002 sec)
```

由于数据库的数据需要升级，现在使用更改命令将原来的数据进行替换，命令如下所示：

```
MySQL [USERDB]> update company set name='baidu' where id=1;
Query OK, 1 row affected (0.036 sec)
Rows matched: 1  Changed: 1  Warnings: 0
MySQL [USERDB]> select * from company;
+----+-------+----------+
| id | name  | addr     |
+----+-------+----------+
|  1 | baidu | china    |
|  2 | TX    | shenzhen |
+----+-------+----------+
2 rows in set (0.002 sec)
```

（8）验证 Mycat 服务对数据库读写操作分离。在 Mycat 虚拟机节点使用 mysql 命令，通过 9066 端口查询对数据库读写操作的分离信息。命令如下所示：

```
[root@mycat ~]# mysql -h 127.0.0.1 -P 9066 -uroot -p000000 -e 'show @@datasource;'
```

查询结果如图 3-3 所示。

```
[root@mycat ~]# mysql -h 127.0.0.1 -P 9066 -uroot -p000000 -e 'show @@datasource;'
+----------+--------+-------+----------------+------+-----+--------+------+------+---------+-----------+------------+
| DATANODE | NAME   | TYPE  | HOST           | PORT | W/R | ACTIVE | IDLE | SIZE | EXECUTE | READ_LOAD | WRITE_LOAD |
+----------+--------+-------+----------------+------+-----+--------+------+------+---------+-----------+------------+
| dn1      | hostM1 | mysql | 192.168.200.14 | 3306 | W   | 0      | 10   | 1000 | 707     | 0         | 7          |
| dn1      | hostS1 | mysql | 192.168.200.15 | 3306 | R   | 0      | 8    | 1000 | 711     | 14        | 0          |
+----------+--------+-------+----------------+------+-----+--------+------+------+---------+-----------+------------+
```

图 3-3　查询 Mycat 服务端口

可以看到所有的写入操作 WRITE_LOAD 数都在 dn1 主数据库节点上，所有的读取操作 READ_LOAD 数都在 dn2 主数据库节点上。由此可见，数据库读写操作已经分离到 dn1 和 dn2 节点上了。

由此，构建读写分离数据库集群任务完成，本实验到此结束。

小　　结

在本章中，您已经学会：
- 数据库服务的运维。
- 主从数据库服务部署。
- 读写分离服务部署。
- 读写分离数据库集群部署。

第 4 章

NoSQL 数据库服务

本章概要

学习 NoSQL 数据库服务的理论知识，掌握 Memcached 和 Redis 服务的部署，掌握 Sentinel 实现 Redis 集群高可用部署。

学习目标

- 了解 NoSQL 数据库服务。
- 掌握 Memcached 服务运维和部署。
- 掌握 Redis 服务运维和部署。
- 掌握 Sentinel 实现 Redis 集群高可用部署。

思维导图

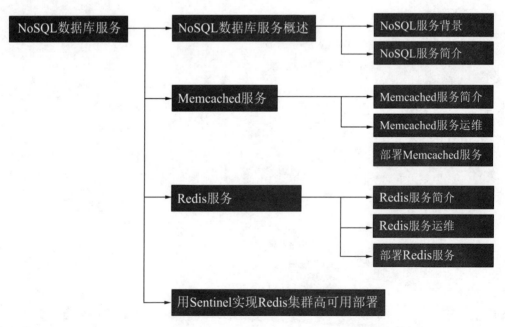

任务目标

学习 Memcached 和 Redis 服务的部署，使用 Sentinel 实现 Redis 集群高可用部署。

4.1 NoSQL 数据库服务概述

学习目标

学完本节后，您应能够：
- 了解 NoSQL 服务背景。
- 了解 NoSQL 服务。

4.1.1 NoSQL 服务背景

1. NoSQL 简史

NoSQL 一词最早出现于 1998 年，是 Carlo Strozzi 开发的一个轻量、开源、不提供 SQL 功能的关系数据库。

2009 年，Last.fm 的 Johan Oskarsson 发起了一次关于分布式开源数据库的讨论，来自 Rackspace 的 Eric Evans 再次提出了 NoSQL 的概念，这时的 NoSQL 主要指非关系型、分布式、不提供 ACID 的数据库设计模式。

2009 年，在亚特兰大举行的"no:sql(east)"讨论会是一个里程碑，其口号是"select fun, profit from real_world where relational=false;"。因此，对 NoSQL 最普遍的解释是"非关联型的"，强调 Key-Value Stores 和文档数据库的优点，而不是单纯的反对 RDBMS。

2. 使用 NoSQL 的原因

用户的个人信息、社交网络、地理位置，用户生成的数据和用户操作日志已经成倍地增加。如果要对这些用户数据进行挖掘，SQL 数据库已经不适用了，NoSQL 数据库的发展却能很好地处理这些数据。

3. 关系型数据库遵循 ACID 规则

事务在英文中是 transaction，和现实世界中的交易很类似，它有如下四个特性：

（1）A（Atomicity）原子性。原子性很容易理解，也就是说事务里的所有操作要么全部做完，要么都不做，事务成功的条件是事务里的所有操作都成功，只要有一个操作失败，整个事务就失败，需要回滚。

（2）C（Consistency）一致性。一致性也比较容易理解，也就是说数据库要一直处于一致的状态，事务的运行不会改变数据库原本的一致性约束。

（3）I（Isolation）独立性。独立性是指并发的事务之间不会互相影响，如果一个事务要访问的数据正在被另外一个事务修改，只要另外一个事务未提交，它所访问的数据就不受未提交事务的影响。

（4）D（Durability）持久性。持久性是指一旦事务提交后，它所做的修改将会永久地保存在数据库上，即使出现宕机也不会丢失。

4.1.2 NoSQL 服务简介

1. NoSQL 简介

数据是当今世界最有价值的资产之一。在大数据时代，人们生产、收集数据的能力大大提升，但是传统的关系型数据库在可扩展性、数据模型和可用性方面已远远不能满足当前的数据处理需求，因此，各种 NoSQL 数据库系统应运而生。

NoSQL 数据库不像关系型数据库那样都有相同的特点，遵循相同的标准。NoSQL 数据库类型多样，可满足不同场景的应用需求，因此取得了巨大的成功。

NoSQL 数据库基本理念是以牺牲事务机制和强一致性机制，来获取更好的分布式部署能力和横向扩展能力，创造出新的数据模型，使其在不同的应用场景下，对特定业务数据具有更强的处理性能。

NoSQL 数据库最初是用于互联网业务需求，互联网数据具有大量化、多样化、快速化等特点。在信息化时代背景下，互联网数据增长迅猛，数据集合规模已实现从 GB、PB 到 ZB 的飞跃。数据不仅是传统的结构化数据，还包含了大量非结构化和半结构化数据，关系型数据库无法存储此类数据。

2. NoSQL 数据库的优势

（1）灵活的数据模型。关系型数据库的数据模型定义严格，无法快速容纳新的数据类型。例如，若要存储客户的电话号码、姓名、地址、城市等信息，则 SQL 数据库需要提前知晓要存储的内容。这对于敏捷开发模式来说十分不方便，因为每次完成新特性时，通常都需要改变数据库的模式。NoSQL 数据库提供的数据模型则能很好地满足这种需求，各种应用可以通过这种灵活的数据模型存储数据而无须修改表；或者只需增加更多的列，无须进行数据的迁移。

（2）可伸缩性强。对企业来说，关系型数据库一开始是普遍的选择。然而，在使用关系型数据库的过程中却遇到了越来越多的问题，原因在于它们是中心化的，是纵向扩展而不是横向扩展的。这使得它们不适合那些需要简单且动态可伸缩性的应用。NoSQL 数据库从一开始就是分布式、横向扩展的，因此非常适合互联网应用分布式的特性。

（3）自动分片。由于关系型数据库存储的是结构化的数据，所以通常采用纵向扩展，即单台服务器要持有整个数据库来确保可靠性与数据的持续可用性。这样做的代价是非常昂贵的，而且扩展也会受到限制。针对这种问题的解决方案就是横向扩展，即添加服务器而不是扩展单台服务器的处理能力。NoSQL 数据库通常支持自动分片，这意味着它们会自动地在多台服务器上分发数据，而不需要应用程序增加额外的操作。

（4）自动复制。NoSQL 数据库支持自动复制。在 NoSQL 数据库分布式集群中，服务器会自动对数据进行备份，即将一份数据复制存储在多台服务器上。因此，当多个用户访问同一数据时，可以将用户请求分散到多台服务器中。同时，当某台服务器出现故障时，其他服务器的数据可以提供备份，即 NoSQL 数据库的分布式集群具有高可用性与灾备恢复的能力。

4.2 Memcached 服务

学习目标

学完本节后，您应能够：
- 了解 Memcached 服务。
- 掌握 Memcached 服务部署。

4.2.1 Memcached 服务简介

Memcached 是 LiveJournal 旗下 Danga Interactive 公司的布拉德·菲茨帕特里克（Brad Fitzpatric）开发的一款内存数据库，现在已被应用于 Facebook、LiveJournal 等公司用于提高 Web 服务质量。

目前这款软件流行于全球各地，经常被用来建立缓存项目，并以此分担来自传统数据库的并发负载压力。

Memcached 可以轻松应对大量同时出现的数据请求，而且它拥有独特的网络结构，在工作机制方面，它还可以在内存中单独开辟新的空间，建立 HashTable，并对 HashTable 进行有效的管理。

1. 使用 Memcached 的原因

由于网站的高并发读写和对海量数据的处理需求，传统的关系型数据库开始出现瓶颈：

（1）对数据库的高并发读写。关系型数据库本身就是个庞然大物，处理过程非常耗时（如解析 SQL 语句、事务处理等）。如果对关系型数据库进行高并发读写（每秒上万次的访问），数据库系统是无法承受的。

（2）对海量数据的处理。对于大型的 SNS 网站（如 Twitter、新浪微博），每天有上千万条的数据产生。对关系型数据库而言，如果在一个有上亿条数据的数据表中查找某条记录，效率将非常低。

使用 Memcached 的主要目的是通过自身内存中缓存关系型数据库的查询结果，减少数据库自身被访问的次数，以提高动态 Web 应用的速度，增强网站架构的并发能力和可扩展性。通过在事先规划好的系统内存空间中临时缓存数据库中的各类数据，以达到减少前端业务服务对关系型数据库的直接高并发访问，从而达到提升大规模网站集群中动态服务的并发访问能力。

（3）Memcached 流程图。Web 服务器读取数据时先读 Memcached 服务器，若 Memcached 没有所需的数据，则向数据库请求数据，然后 Web 再把请求到的数据发送到 Memcached，如图 4-1 所示。

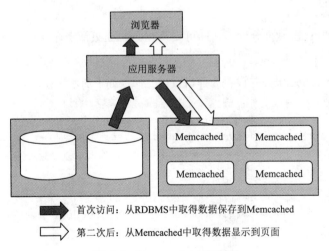

图 4-1 Memcached 流程图

2. Memcached 的特征

（1）协议简单。Memcached 的服务器客户端通信并不使用复杂的 .xm 等格式，而是使用简单的基于文本行的协议。

（2）基于 libevent 的事件处理。libevent 是个程序库，它将 Linux 的 epoll、BSD 类操作系统的 kqueue 等事件处理功能封装成统一的接口，即使对服务器的连接数增加，也能发挥 O(1) 的性能。Memcached 使用这个 libevent 库，因此可以在 Linux、BSD、Solaris 等操作系统上发挥其高性能。

（3）采用内置内存存储方式。为了提高性能，Memcached 中保存的数据都存储在 Memcached 内置的内存存储空间中。由于数据仅存在于内存中，所以重启 Memcached 或操作系统都会导致全部数据消失。

（4）不互相通信的分布式架构。Memcached 尽管是"分布式"缓存服务器，但服务器端并没有分布式功能。各个 Memcached 不会互相通信以共享信息。

如图 4-2 所示为 Memcached 分布式架构。

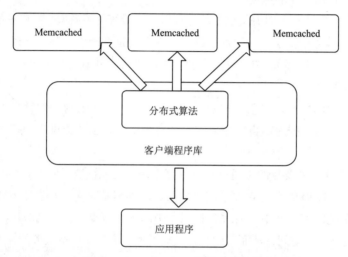

图 4-2 Memcached 分布式架构图

4.2.2 Memcached 服务运维

1. 服务安装

（1）在配置妥当 yum 软件仓库后，即可安装 Memcached 服务了。命令如下所示：

```
[root@memcache ~]# yum install memcached libmemcached libevent nc
```

（2）启动 Memcached 服务并设置为开机自启。命令如下所示：

```
[root@memcache ~]# systemctl start memcached
[root@memcache ~]# systemctl enable memcached
ln -s '/usr/lib/systemd/system/memcached.service' '/etc/systemd/system/multi-user.target.wants/memcached.service'
```

（3）查看启动情况：

```
[root@memcache ~]# ps ax | grep memcached
11409 ?        Ssl    0:00 /usr/bin/memcached -u memcached -p 11211 -m 64 -c 1024
11425 pts/0    S+     0:00 grep --color=auto memcached
```

2. 服务运维

（1）memcached-tool 是用 perl 语言写的一个对 Memcache 的状态性能分析分具，使用命令查看服务当前数据信息。

```
[root@memcache ~]# memcached-tool 127.0.0.1:11211
```

```
 # Item_Size    Max_age    Pages    Count    Full?    Evicted Evict_Time OOM
```

（2）使用命令向 Memcached 中添加数据时，注意添加的数据一般为键值对的形式。向 Memcached 中添加数据时，注意添加的数据一般为键值对的形式，例如：key1-->values1,key2-->values2。这里把 Memcached 添加、查询、删除等的命令和 MySQL 数据库做一个基本类比，如图 4-3 所示。

MySQL数据库管理	Memcached管理
MySQL的insert语句	Memcached的set命令
MySQL的select语句	Memcached的get命令
MySQL的delete语句	Memcached的delete命令

图 4-3　Memcached 管理命令

通过 printf 配合 nc 向 Memcached 中写入数据，然后检查命令如下所示：

```
[root@memcache ~]# printf "set key1 0 0 5\r\nbenet\r\n" | nc 127.0.0.1 11211
STORED
[root@memcache ~]# printf "get key1\r\n" | nc 127.0.0.1 11211
VALUE key1 0 5
benet
END
```

printf 和 echo 的区别是 printf 不会输出换行符，nc 是用于连接 Memcached 的。

（3）查看数据：

```
[root@memcache ~]# memcached-tool 127.0.0.1:11211
 # Item_Size    Max_age    Pages    Count    Full?    Evicted Evict_Time OOM
 1  96B        80s         1        1        yes      0       0          0
```

查看 Memcached 运行状态：

```
[root@memcache ~]#  memcached-tool 127.0.0.1:11211 stats
#127.0.0.1:11211    Field           Value
             accepting_conns        1
             auth_cmds              0
             auth_errors            0
             bytes                  74
             bytes_read             99
             bytes_written          1171
             cas_badval             0
             cas_hits               0
             cas_misses             0
             cmd_flush              0
             cmd_get                1
             cmd_set                1
             cmd_touch              0
             conn_yields            0
      connection_structures         11
             curr_connections       10
             curr_items             1
             decr_hits              0
             decr_misses            0
             delete_hits            0
             delete_misses          0
```

```
            evicted_unfetched          0
                    evictions          0
            expired_unfetched          0
                     get_hits          1
                   get_misses          0
                   hash_bytes          524288
            hash_is_expanding          0
             hash_power_level          16
                    incr_hits          0
                  incr_misses          0
                     libevent          2.0.21-stable
                limit_maxbytes         67108864
           listen_disabled_num         0
                          pid          11409
                 pointer_size          64
                    reclaimed          0
                 reserved_fds          20
                rusage_system          0.019748
                  rusage_user          0.023451
                      threads          4
                         time          1402913371
            total_connections          16
                  total_items          1
                   touch_hits          0
                 touch_misses          0
                       uptime          831
                      version          1.4.15
```

（4）通过 printf 配合 nc 从 Memcached 中删除数据。

```
[root@memcache ~]# printf "delete key1\r\n" | nc 127.0.0.1 11211
DELETED
[root@memcache ~]# printf "get key1\r\n" | nc 127.0.0.1 11211
END
```

推荐使用上述方法测试操作 Memcached，只支持键值的读取方式，别的公式不支持。

4.2.3　任务：部署 Memcached 服务

Memcached 是一个高性能的分布式内存对象缓存系统，用于动态 Web 应用以减轻数据库负载。

任务执行清单

在本任务中，您将掌握 Memcached 服务的部署。

目标

- 掌握 Memcached 服务的部署。

重要信息

- 提前配置 yum 源文件。
- 提前搭建 PHP 环境配置。

解决方案

1. 服务安装

（1）在 PHP 环境中，若希望让 PHP 连接 Memcached 服务，需要安装 PHP 的 Memcached 扩

展模块 memcache，将 memcache-3.0.8.tgz 上传到 root 目录下，并解压编译安装，命令如下所示：

```
[root@lnmp ~]# tar -zxf memcache-3.0.8.tgz
[root@lnmp ~]# cd memcache-3.0.8
[root@lnmp memcache-3.0.8]# /usr/local/php5.6/bin/php
php          php-cgi     php-config    phpize
[root@lnmp memcache-3.0.8]# /usr/local/php5.6/bin/phpize
Configuring for:
PHP Api Version:          20131106
Zend Module Api No:       20131226
Zend Extension Api No:    220131226
[root@lnmp memcache-3.0.8]# ./configure --with-php-config=/usr/local/php5.6/bin/php-config
[root@lnmp memcache-3.0.8]# make && make install
```

（2）在 PHP 中增加缓存模块，并重新加载，编辑 php.ini 配置文件，在代码 ";extension=php_shmop.dll"下面添加 extension=memcache.so，命令如下所示：

```
[root@lnmp memcache-3.0.8]# vi /etc/php.ini
;extension=php_shmop.dll
extension=memcache.so
```

（3）使用 /usr/local/php5.6/bin/php -m 命令可以查看是否成功安装了 Memcache 扩展模块，然后重启 PHP 服务，命令如下所示：

```
[root@lnmp memcache-3.0.8]#  /usr/local/php5.6/bin/php -m|grep memcache
memcache
[root@lnmp memcache-3.0.8]# systemctl restart php-fpm
```

2. 实验测试

（1）将测试的两个文件复制到相应的启动文件夹中，然后更改配置文件的信息，使得与本机服务配置同步，最后启动 Memcached 服务，命令和参数如下所示：

```
[root@lnmp memcache-3.0.8]# cp memcache.php /usr/local/nginx/html/
[root@lnmp memcache-3.0.8]# cp example.php /usr/local/nginx/html/
[root@lnmp memcache-3.0.8]# cd /usr/local/nginx/html/
[root@lnmp html]# vi memcache.php
$MEMCACHE_SERVERS[] = '192.168.200.10:11211'; // add more as an array
[root@lnmp memcache-3.0.8]# systemctl start memcached
```

（2）打开浏览器，输入"IP/memcache.php"，打开 php 服务测试 Memcached 服务的网页，其中 Hits 命中率为 50%，如图 4-4 所示。

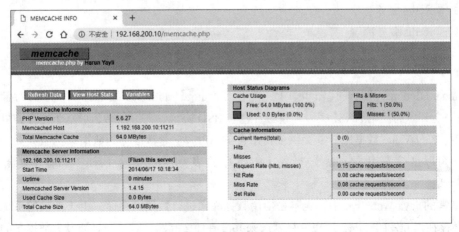

图 4-4　网页服务测试

（3）输入"IP/example.php"进行网页的测试，此时不断刷新此网页之后，如图 4-5 所示，回到上一步，监控页面的数据会跟着改变，命中率不断提高，最终 Hits 命令达到 100%，如图 4-6 所示，至此，Memcached 服务配置成功。

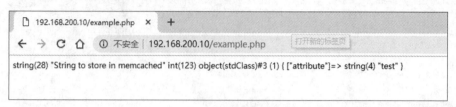

图 4-5　PHP 网页测试

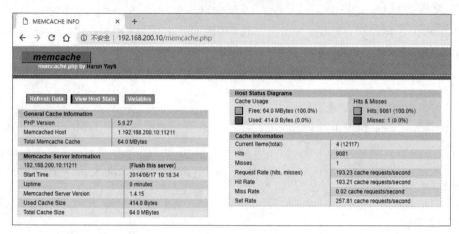

图 4-6　测试成功

本任务到此结束。

4.3　Redis 服务

学习目标

学完本节后，您应能够：
- 了解 Redis 服务。
- 掌握 Redis 服务部署。

4.3.1　Redis 服务简介

REmote DIctionary Server（Redis）是一个由 Salvatore Sanfilippo 编写的 key-value 存储系统。Redis 是一个开源的使用 ANSI C 语言编写、遵守 BSD 协议、支持网络、可基于内存亦可持久化的日志型、Key-Value 数据库，并提供多种语言的 API。它通常被称为数据结构服务器，因为值（value）可以是 字符串（String）、哈希（Hash）、列表（list）、集合（sets）和有序集合（sorted sets）等类型。

在日常的 Java Web 开发中，无不是使用数据库来进行数据的存储，由于一般的系统任务中通常不会存在高并发的情况，所以这样看起来并没有什么问题，可是一旦涉及大数据量的需求，例如一些商品抢购的情景,或者是主页访问量瞬间增大的时候,单一使用数据库来保存数据的系统会因为面向磁盘,磁盘读/写速度比较慢的问题而存在严重的性能弊端,一瞬间成千上万的请求到来,需要系统在极短的时间内完成成千上万次的读/写操作,这个时候往往不是数据库能够承受的,极容易造成数据库系统瘫痪,最终导致服务宕机的严重问题。

1. Sentinel 的工作方式

Sentinel 是 Redis 官方提供的一种高可用方案(除了 Sentinel,Redis Cluster 是另一种方案),它可以自动监控 Redis master/slave 的运行状态,如果发现 master 无法访问了,就会启动 failover 把其中一台可以访问的 slave 切换为 master。

支持 Sentinel 的 Redis 客户端(例如 Java 的 Jedis)会在连接 Redis 服务器的时候向 Sentinel 询问 master 的 IP,并且会在收到 master 切换的 pub/sub 事件后自动重新连接到新的 master。对调用 Redis 客户端的业务系统来说,这些都是完全透明的。

2. Sentinel 的作用

Sentinel 作用:检测 master 状态,如果 master 异常,则会进行 master-slave 切换,将其中一个 Slave 作为 master,将之前的 master 作为 Slave。当 master-slave 切换后,master-redis.conf、slave-redis.conf 和 sentinel.conf 的内容都会发生改变,即 master-redis.conf 中会多一行 slaveof 的配置,sentinel.conf 的监控目标会随之调换。

3. Redis 与其他 key-value 存储的差异

(1) Redis 有着更为复杂的数据结构并且提供对它们的原子性操作,这是一个不同于其他数据库的进化路径。Redis 的数据类型都是基于基本数据结构的同时对程序员透明,无须进行额外的抽象。

(2) Redis 运行在内存中但是可以持久化到磁盘,所以在对不同数据集进行高速读写时需要权衡内存,因为数据量不能大于硬件内存。在内存数据库方面的另一个优点是,相比在磁盘上相同的复杂的数据结构,在内存中操作起来非常简单,这样 Redis 可以做很多内部复杂性很强的事情。同时,在磁盘格式方面是紧凑的以追加的方式产生的,因为并不需要进行随机访问。

4. Redis 服务的特点

(1) Redis 支持数据的持久化,可以将内存中的数据保存在磁盘中,重启的时候可以再次加载进行使用。

(2) Redis 不仅仅支持简单的 key-value 类型的数据,同时还提供 list、set、zset、hash 等数据结构的存储。

(3) Redis 支持数据的备份,即 master-slave 模式的数据备份。

5. Redis 服务的优势

(1) 性能极高:Redis 能读的速度是 110 000 次/s,写的速度是 81 000 次/s。

(2) 丰富的数据类型:Redis 支持二进制案例的 Strings、Lists、Hashes、Sets 及 Ordered Sets 数据类型操作。

(3) 原子:Redis 的所有操作都是原子性的,意思就是要么成功执行,要么失败完全不执行。单个操作是原子性的。多个操作也支持事务,即原子性,通过 MULTI 和 EXEC 指令包起来。

(4) 丰富的特性:Redis 还支持 publish/subscribe、通知、key 过期等特性。

4.3.2 Redis 服务运维

1. 服务安装

(1) 在配置妥当 yum 软件仓库后,即可安装 Redis 服务。命令如下所示:

```
[root@redis ~]# yum -y install redis
```

(2) 修改 Redis 配置文件,编辑 /etc/redis.conf 文件。将 bind 127.0.0.1 这一行注释掉,将 protected-mode yes 改为 protected-mode no。

```
[root@redis ~]# systemctl start redis
[root@redis ~]# systemctl enable redis
Created symlink from /etc/systemd/system/multi-user.target.wants/redis.service to /usr/lib/systemd/system/redis.service.
```

(3) 检查 Redis 服务启动,如果检查到 6379 端口,即证明 Redis 服务运行成功。命令如下:

```
[root@redis ~]# netstat -ntapl | grep 6379
tcp    0   0  0.0.0.0:6379    0.0.0.0:*    LISTEN    7392/redis-server *
tcp6   0   0  :::6379         :::*         LISTEN    7392/redis-server *
```

2. 服务运维

(1) Redis 命令用于在 Redis 服务上执行操作。要在 Redis 服务上执行命令需要一个 Redis 客户端。Redis 客户端在之前下载的 Redis 的安装包中。

启动 Redis 服务器,打开终端并输入命令 redis-cli,该命令会连接本地的 Redis 服务。

```
[root@redis ~]# redis-cli
127.0.0.1:6379> ping
PONG
```

在以上实例中连接到本地的 Redis 服务并执行 PING 命令,该命令用于检测 Redis 服务是否启动。

(2) Redis 列表是简单的字符串列表,按照插入顺序排序。可以添加一个元素到列表的头部(左边)或者尾部(右边)。一个列表最多可以包含 $2^{32} - 1$ 个元素(4 294 967 295,每个列表超过 40 亿个元素)。

```
127.0.0.1:6379> LPUSH runoobkey redis
(integer) 1
127.0.0.1:6379> LPUSH runoobkey mongodb
(integer) 2
127.0.0.1:6379> LPUSH runoobkey mysql
(integer) 3
127.0.0.1:6379>  LRANGE runoobkey 0 10
1) "mysql"
2) "mongodb"
3) "redis"
```

(3) Redis 的 Set 是 String 类型的无序集合。集合成员是唯一的,这就意味着集合中不能出现重复的数据。Redis 中集合是通过哈希表实现的,所以添加、删除、查找的复杂度都是 O(1)。

集合中最大的成员数为 $2^{32}-1$(4 294 967 295,每个集合可存储 40 多亿个成员)。

```
redis 127.0.0.1:6379> SADD runoobkey redis
(integer) 1
redis 127.0.0.1:6379> SADD runoobkey mongodb
(integer) 1
```

```
redis 127.0.0.1:6379> SADD runoobkey mysql
(integer) 1
redis 127.0.0.1:6379> SADD runoobkey mysql
(integer) 0
redis 127.0.0.1:6379> SMEMBERS runoobkey
1) "mysql"
2) "mongodb"
3) "redis"
```

在以上实例中我们通过 SADD 命令向名为 runoobkey 的集合插入三个元素。

（4）Redis 字符串数据类型的相关命令用于管理 Redis 字符串值。

```
127.0.0.1:6379> SET runoobkey redis
OK
127.0.0.1:6379> GET runoobkey
"redis"
```

（5）Redis 脚本使用 Lua 解释器来执行脚本。Redis 2.6 版本通过内嵌支持 Lua 环境。执行脚本的常用命令为 EVAL。

```
127.0.0.1:6379> EVAL "return {KEYS[1],KEYS[2],ARGV[1],ARGV[2]}" 2 key1 key2 first second
1) "key1"
2) "key2"
3) "first"
4) "second"
```

（6）Redis SAVE 命令用于创建当前数据库的备份。

```
127.0.0.1:6379> save
OK
```

该命令将在 Redis 安装目录中创建 dump.rdb 文件。如果需要恢复数据，只需将备份文件（dump.rdb）移动到 Redis 安装目录并启动服务即可。获取 redis 目录可以使用 CONFIG 命令，如下所示：

```
127.0.0.1:6379> config get dir
1) "dir"
2) "/var/lib/redis"
```

以上命令 CONFIG GET dir 输出的 redis 安装目录为 /usr/local/redis/bin。

4.3.3 任务：部署 Redis 服务

Redis 是一种非关系型数据存储工具，这区别于传统的关系型数据库（如 MySQL 等），类似于 Memcached，并且其内部集成了对 list（链表）、set（集合）的操作，可以很方便快速地处理数据（如插入、删除 list 取交集、并集、差集等），这极大地减轻了底层数据库的压力，并且给用户带来了更快的响应速度。

任务执行清单

在本任务中，您将掌握 Redis 服务的部署。

目标

- 掌握 Redis 服务的部署。

重要信息

- 提前配置 yum 源文件。
- 提前搭建 PHP 的配置。

解决方案

1. 服务安装

（1）下载稳定版本的 Redis，将其上传至 root 目录下，并解压到 /usr/local/src/ 目录下，进入目录进行编译安装。

```
[root@lnmp ~]# tar -zxvf redis-4.0.8.tar.gz -C /usr/local/src/
[root@lnmp ~]# cd /usr/local/src/redis-4.0.8/
[root@lnmp redis-4.0.8]# make && make install
```

（2）将 Redis 的配置文件 redis.conf 复制至 /etc 目录下，命令如下所示：

```
[root@lnmp redis-4.0.8]# cp redis.conf /etc/
cp: overwrite '/etc/redis.conf'? y
```

复制后，修改 /etc/redis.conf 配置文件，修改如下所示。启动 Redis 服务，通过 redis-server 指定 redis.conf 文件启动，默认监听端口是 6379。

```
[root@lnmp redis-4.0.8]# vi /etc/redis.conf
daemonize yes
logfile "/var/log/redis.log"
dir /data/redis
appendonly yes
[root@lnmp redis-4.0.8]# mkdir -p /data/redis
[root@lnmp redis-4.0.8]# redis-server /etc/redis.conf
```

（3）内核警告 WARNING overcommit_memory is set to 0 和 WARNING you have Transparent Huge Pages（THP）support enabled，建议修改如下内容：

```
[root@lnmp redis-4.0.8]# less /var/log/redis.log
[root@lnmp redis-4.0.8]# sysctl vm.overcommit_memory=1
vm.overcommit_memory = 1
[root@lnmp redis-4.0.8]# echo never >
  /sys/kernel/mm/transparent_hugepage/enabled
[root@lnmp redis-4.0.8]# vi /etc/rc.local
```

最后添加至 rc.loal 配置文件中，让服务器启动时，直接执行这两条命令，命令如下所示：

```
echo never > /sys/kernel/mm/transparent_hugepage/enabled
sysctl vm.overcommit_memory=1
```

2. 安装 PHP 模块

（1）将 phpredis-4.1.0.tar.gz 上传至 root 目录下，解压到 /usr/local/src/ 目录下，并编译安装 Redis 扩展模块，命令如下所示：

```
[root@lnmp ~]# tar -zxvf phpredis-4.1.0.tar.gz -C /usr/local/src/
[root@lnmp ~]# cd /usr/local/src/phpredis-4.1.0/
[root@lnmp phpredis-4.1.0]# /usr/local/php5.6/bin/phpize
Configuring for:
PHP Api Version:         20131106
Zend Module Api No:      20131226
```

```
Zend Extension Api No:    220131226
[root@lnmp phpredis-4.1.0]# ./configure --with-php-config=/usr/local/php5.6/bin/php-config
[root@lnmp phpredis-4.1.0]# make && make install
```

(2)在 PHP 中增加缓存模块,并重新加载,编辑 php.ini 配置文件,在代码 ";extension=php_shmop.dll" 下面添加 extension=redis.so,命令如下所示:

```
[root@lnmp phpredis-4.1.0]# vi /etc/php.ini
extension=redis.so
```

(3)使用 /usr/local/php5.6/bin/php -m 命令可以查看是否成功安装了 redis 扩展模块,命令如下所示,最后重新启动 PHP 服务。

```
[root@lnmp phpredis-4.1.0]# /usr/local/php5.6/bin/php -m|grep redis
redis
[root@lnmp phpredis-4.1.0]# systemctl restart php-fpm
```

3. 服务测试

(1)登录数据库将测试文件导入 test 数据库中,然后查看数据库,命令如下所示:

```
[root@lnmp ~]# mysql -uroot -p000000
Welcome to the MariaDB monitor.  Commands end with ; or \g.
Your MariaDB connection id is 5005
Server version: 5.5.56-MariaDB MariaDB Server
Copyright (c) 2000, 2017, Oracle, MariaDB Corporation Ab and others.
Type 'help;' or '\h' for help. Type '\c' to clear the current input statement.
MariaDB [(none)]> use test;
Database changed
MariaDB [test]> source /root/test.sql
Query OK, 0 rows affected (0.00 sec)
Query OK, 0 rows affected (0.00 sec)
Query OK, 0 rows affected (0.00 sec)
Query OK, 0 rows affected (0.00 sec)
Query OK, 0 rows affected (0.00 sec)
Query OK, 0 rows affected (0.00 sec)
Query OK, 0 rows affected (0.00 sec)
Query OK, 0 rows affected (0.00 sec)
Query OK, 0 rows affected (0.00 sec)
Query OK, 0 rows affected (0.00 sec)
Query OK, 0 rows affected (0.00 sec)
Query OK, 0 rows affected (0.00 sec)
Query OK, 0 rows affected (0.00 sec)
Query OK, 0 rows affected (0.12 sec)
Query OK, 0 rows affected (0.00 sec)
Query OK, 0 rows affected (0.00 sec)
Query OK, 0 rows affected (0.01 sec)
Query OK, 9 rows affected (0.01 sec)
Records: 9  Duplicates: 0  Warnings: 0
MariaDB [test]> show tables;
+----------------+
| Tables_in_test |
```

```
+----------------+
| test           |
+----------------+
1 row in set (0.00 sec)
MariaDB [test]> select * from test;
+----+-------+
| id | name  |
+----+-------+
|  1 | test1 |
|  2 | test2 |
|  3 | test3 |
|  4 | test4 |
|  5 | test5 |
|  6 | test6 |
|  7 | test7 |
|  8 | test8 |
|  9 | test9 |
+----+-------+
9 rows in set (0.00 sec)
MariaDB [test]> exit
Bye
```

（2）编写测试页面，具体代码在 index.php 文件中，如图 4-7 所示。

```php
[root@lnmp ~]# vi /usr/local/nginx/html/index.php
<?php
    $redis = new Redis();
    $redis->connect('127.0.0.1',6379) or die("Could not connect redis server");
    $query = "select * from test";
    for($key = 1;$key < 10;$key++)
    {
        if(!$redis->get($key))
        {
            $connect = mysql_connect('127.0.0.1','root','000000');
            mysql_select_db(test);
            $result = mysql_query($query);
            //如果没有找到$key,就将该查询sql的结果缓存到redis
            while($row = mysql_fetch_assoc($result))
            {
                $redis->set($row['id'],$row['name']);
            }
            $myserver = 'mysql';
            break;
        }
        else
        {
            $myserver = "redis";
            $data[$key] = $redis->get($key);
        }
    }
    echo $myserver;
    echo "<br>";
    for($key = 1;$key < 10;$key++){
        echo "number is <b><font color=#FF0000>$key</font></b>";
        echo "<br>";
        echo "name is <b><font color=#FF0000>$data[$key]</font></b>";
        echo "<br>";
    }
?>
```

图 4-7 配置文件

（3）网页测试，在浏览器中输入 192.168.200.10/index.php，即可通过访问 redis 缓存，进而访问数据库表信息，如图 4-8 所示。

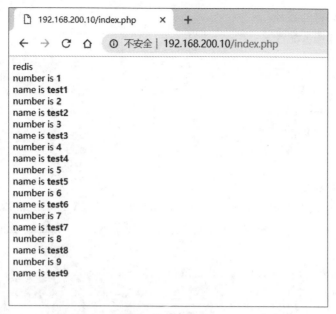

图 4-8 网页测试

```
[root@lnmp ~]# redis-cli
127.0.0.1:6379> get
(error) ERR wrong number of arguments for 'get' command
127.0.0.1:6379> get 1
"test1"
127.0.0.1:6379> get 2
"test2"
127.0.0.1:6379> get 3
"test3"
```

本任务到此结束。

4.4 开放研究任务：用 Sentinel 实现 Redis 集群高可用部署

公司的 Redist 主从架构中，如果主服务器离线，那么所有写操作无法执行。为了避免此情况发生，Redis 引入了 Sentinel（哨兵）机制，当 Redis 主出现故障了，Redis 从可以主动变成主。

任务执行清单

在本任务中，您将通过实践使用 Sentinel 实现 Redis 集群高可用部署。
（1）配置基础环境。
（2）配置主从 Redis 架构。
（3）Sentinel 服务配置。
（4）服务测试。

目标

- 掌握 Redis 服务的部署。

- 掌握 Redis 集群的部署。
- 掌握 Sentinel 服务配置。

重要信息

- 准备 3 台全新服务器。
- 提前配置好 yum 源文件。
- 虚拟机密码为 000000。

解决方案

1. 配置基础环境

（1）使用远程连接工具 CRT 连接到 192.168.200.11、192.168.200.12、192.168.200.13 这三台服务器，并对三台服务器进行修改主机名的操作，分别为 redis1、redis2、redis3。此处以 redis1 主节点为例。

```
[root@localhost ~]# hostnamectl set-hostname redis1
[root@localhost ~]# logout
[root@redis1 ~]# hostnamectl
   Static hostname: redis1
         Icon name: computer-vm
           Chassis: vm
        Machine ID: 622ba110a69e24eda2dca57e4d306baa
           Boot ID: 3133e7777c0942fa8a8c2ac486998554
    Virtualization: kvm
  Operating System: CentOS Linux 7 (Core)
       CPE OS Name: cpe:/o:centos:centos:7
            Kernel: Linux 3.10.0-862.2.3.el7.x86_64
      Architecture: x86-64
```

（2）登录到服务器，使用提前配置好的 yum 源文件，分别安装 Redis 服务，如果没有配置好，可以通过以下方式配置 yum 仓库地址。此处以 redis1 主节点为例。

```
[root@redis1 ~]# yum -y install epel-release
[root@redis1 ~]# yum -y install http://rpms.remirepo.net/enterprise/remi-release-7.rpm
[root@redis1 ~]# yum -y install redis
```

2. 配置主从 Redis 架构

（1）修改 redis1 主机上的 Redis 配置文件，然后启动服务，并使用命令查看 6379 端口是否开启，命令如下所示：

```
[root@redis1 ~]# vi /etc/redis.conf
改: bind 127.0.0.1
为: bind 0.0.0.0           #redis监听的地址，改为0.0.0.0表示在所有网卡接口上进行监听。
改: protected-mode yes
为:  protected-mode no    # 关闭protected-mode,允许外网访问redis服务器。
[root@redis1 ~]# systemctl restart redis
[root@redis1 ~]# netstat -ntapl | grep 6379
tcp    0    0 0.0.0.0:6379        0.0.0.0:*          LISTEN        10966/redis-server
```

（2）修改 redis2 和 redis3 主机上的 Redis 配置文件，然后启动服务，此处以 redis2 为例，命令如下所示：

```
[root@redis2 ~]# vi /etc/redis.conf
```

```
改: bind 127.0.0.1
为: bind 0.0.0.0  #redis 监听的地址，改为 0.0.0.0 表示在所有网卡接口上进行监听。
改: protected-mode yes
为: protected-mode no
改: # slaveof <masterip> <masterport>
为: slaveof 192.168.200.11 6379
[root@redis2 ~]# systemctl restart redis
```

（3）登录到从节点 redis2 主机服务器上，使用命令查看主从复制状态，命令如下所示：

```
[root@redis2 ~]# redis-cli
127.0.0.1:6379> info replication
# Replication
role:slave                          # 角色: slave
master_host:192.168.200.11          # 主服务器 IP
master_port:6379                    # 主服务器端口
master_link_status:up               # 主服务器连接状态为 up，说明已经主从同步上了
...
```

至此，一主二从的 Redis 主从复制架构已经搭建成功，下面开始配置 Sentinel 架构。

3. Sentinel 服务配置

（1）登录到 redis1 节点的主服务器上，配置 Sentinel 服务配置文件，命令如下所示：

```
[root@redis1 ~]# vi /etc/redis-sentinel.conf
改: # protected-mode no
为: protected-mode no
改: sentinel monitor mymaster 127.0.0.1 6379 2
为: sentinel monitor mymaster 1921.68.200.11 6379 2
改: sentinel down-after-milliseconds mymaster 30000  #默认单位是毫秒，配置 10 秒
为: sentinel down-after-milliseconds mymaster 10000
改: sentinel failover-timeout mymaster 180000
为: sentinel failover-timeout mymaster 60000
```

sentinel monitor <master-name> <ip> <redis-port> <quorum> 每项含意如下：

<master-name> 为集群名称，可以自定义，如果同时监控多组 Redis 集群时，<master-name> 不能一样。

<ip> 为主节点的 IP 地址。

<redis-port> 为主节点的端口号。

<quorum> 主节点对应的 quorum 法定数量，用于定义 Sentinel 的数量，是一个大于值，尽量使用奇数，如果 Sentinel 有 3 个，则指定为 2 即可，如果有 4 个，不能够指定为 2，避免导致集群分裂。最后的数字 2 指定的值，表明如果有 2 个 Sentinels 无法连接 master，才认为 master 故障了。

（2）将此配置文件分别在 redis2 和 redis3 中配置，此处使用命令直接发送过去，命令如下所示：

```
[root@redis1~]# scp /etc/redis-sentinel.conf 192.168.200.12:/etc/redis-sentinel.conf
[root@redis1~]# scp /etc/redis-sentinel.conf 192.168.200.13:/etc/redis-sentinel.conf
```

（3）复制完成之后，在三台服务器中分别启动 Redis 服务和 Sentinel 服务器。命令如下所示：

```
[root@redis1 ~]# systemctl start redis && systemctl start redis-sentinel
[root@redis2 ~]# systemctl start redis && systemctl start redis-sentinel
[root@redis3 ~]# systemctl start redis && systemctl start redis-sentinel
```

4. 服务测试

（1）首先在 redis1 节点中查看当前主从的服务状态，命令和结果如下所示：

```
[root@redis1 ~]# redis-cli -h 192.168.200.13
192.168.200.13:6379> info replication
# Replication
role:slave                                    # 服务的权限是 slave
master_host:192.168.200.11                    # 主服务器的 IP
master_port:6379
master_link_status:up
```

（2）模拟故障，现在将 redis1 主节点服务停止，然后使用刚才使用的命令查看服务状态信息，命令和结果如下所示：

```
[root@redis1 ~]# systemctl stop redis
[root@redis1 ~]# redis-cli -h 192.168.200.13
192.168.200.13:6379> info replication
# Replication
role:master                                   # 服务的权限变成了 master
connected_slaves:1                            # 连接的从节点变成了 1 个
slave0:ip=192.168.200.12,port=6379,state=online,offset=7358,lag=0
master_repl_offset:7358
```

说明此时 redis3 节点 192.168.200.13 的服务器变成主节点了。

（3）重新启动服务，恢复 redis1 服务的状态，继续使用命令查看，主节点依然是 redis3，不会因为 redis1 的恢复变更，命令和结果如下所示：

```
[root@redis1 ~]# systemctl start redis
[root@redis1 ~]# redis-cli -h 192.168.200.13
192.168.200.13:6379> info replication
# Replication
role:master                                   # 权限为 master
connected_slaves:2                            # 连接的从节点数为 2 个
slave0:ip=192.168.200.12,port=6379,state=online,offset=32407,lag=1
slave1:ip=192.168.200.11,port=6379,state=online,offset=32407,lag=1
# 将 192.168.200.11 节点变成了从节点
[root@redis2 ~]# redis-cli
127.0.0.1:6379> info replication
# Replication
role:slave
master_host:192.168.200.13
master_port:6379
master_link_status:up
```

主 redis 不会因为 redis1 恢复成功后，就主动让出权限。这样可以避免再次回切时，发生服务中断。

（4）登录 redis3 服务器，查看主从服务器的状态信息，命令和结果如下所示：

```
[root@redis3 ~]# redis-cli
127.0.0.1:6379> info replication
# Replication
role:master
connected_slaves:2
slave0:ip=192.168.200.12,port=6379,state=online,offset=74176,lag=0
slave1:ip=192.168.200.11,port=6379,state=online,offset=74176,lag=0
```

（5）在服务器中查看 Sentinel 服务的状态信息。命令和结果如下所示：

```
[root@redis1 ~]# redis-cli -h 192.168.200.11 -p 26379
192.168.200.11:26379> info sentinel
# Sentinel
sentinel_masters:1
sentinel_tilt:0
sentinel_running_scripts:0
sentinel_scripts_queue_length:0
sentinel_simulate_failure_flags:0
master0:name=mymaster,status=ok,address=192.168.200.13:6379,slaves=2,sentinels=3
[root@redis2 ~]# redis-cli -h 192.168.200.12 -p 26379
192.168.200.12:26379> info sentinel
# Sentinel
sentinel_masters:1
sentinel_tilt:0
sentinel_running_scripts:0
sentinel_scripts_queue_length:0
sentinel_simulate_failure_flags:0
master0:name=mymaster,status=ok,address=192.168.200.13:6379,slaves=2,sentinels=3
```

将来客户端应连接 Sentinel，向 Sentinel 发请求去寻址，并根据 Sentinel 的反馈，连接新的 Redis 主节点，这一点需要使用 Redis 专用客户端来实现。Redis 客户端会根据 Sentinel 返回的新节点 IP 进行连接。这就是开发人员的工作了。在 Java API 中，JedisClient 就提供了使用 Sentinel 的功能。

小　　结

在本章中，您已经学会：

- NoSQL 数据库服务知识。
- Memcached 服务运维和部署。
- Redis 服务运维和部署。
- Sentinel 实现 Redis 集群高可用部署。

第 5 章

共享存储与分布式存储

📝 本章概要

学习共享存储与分布式存储的服务知识，掌握 NFS 共享存储及分布式存储的部署。

🎯 学习目标

- 学习共享存储和分布式存储。
- 掌握 NFS 共享存储部署。
- 掌握 GlusterFS 存储部署。
- 掌握 Ceph 存储部署。
- 掌握 Swift 存储部署。
- 掌握 Ceph 分布式存储集群部署。

🧠 思维导图

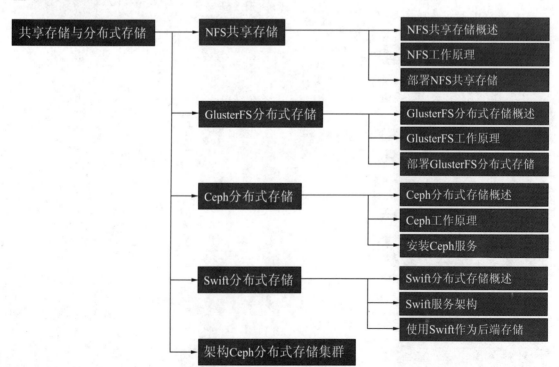

第 5 章 共享存储与分布式存储

任务目标

学习共享存储和分布式存储服务的部署，使用 Ceph 实现分布式存储集群高可用部署。

5.1 NFS 共享存储

学习目标

学完本节后，您应能够：

- 了解什么是 NFS。
- 了解 NFS 工作原理。
- 掌握 NFS 共享存储服务部署。

5.1.1 NFS 共享存储概述

NFS 是 Network File System 的缩写，中文意思是网络文件系统。它的主要功能是通过网络（一般是局域网）让不同的主机系统之间可以共享文件或目录。NFS 客户端（一般为应用服务器，如 Web）可以通过挂载（mount）的方式将 NFS 服务器端共享的数据目录挂载在 NFS 客户端本地系统中（就是某一个挂载点下）。从客户端本地看，NFS 服务器端共享的目录就好像是客户端自己的磁盘分区或目录一样，而实际上确是远端的 NFS 服务器的目录。

NFS 网络文件系统很像 Windows 系统的网络共享、安全功能、网络驱动器映射，这也和 Linux 系统里的 samba 服务类似。只不过一般情况下，Windows 网络共享服务或 samba 服务用于办公局域网共享，而互联网中小型网站集群架构后端常用 NFS 进行数据共享，如果是大型网站，那么有可能还会用到更复杂的分布式文件系统，如 Moosefs(mfs)、GlusterFS、FastDFS。

NFS 系统已经历经了 30 年的发展，是一个非常稳定的（可移植）网络文件系统。在中小型互联网企业中，应用十分广泛。

5.1.2 NFS 工作原理

1. NFS 共享与客户端挂载

在 NFS 服务器端设置好一个共享目录 /data 后，其他有权限访问 NFS 服务器端的客户端都可以将这个共享目录 /data 挂载到客户端本地的某个挂载点（其实就是一个目录，这个挂载点目录可以自己随意指定），不同客户端的挂载点可以不相同。

客户端正确挂载完毕后，就进入了 nfs 客户端的挂载点所在的 /opt/data 或 /data 目录，此时就可以看到 NFS 服务器端 /data 共享出来的目录下的所有数据。在客户端上查看时，NFS 服务器端的 /data 目录就相当于客户端本地的磁盘分区或目录，几乎感觉不到使用上的区别。

已经挂载完后用 df -h 可以看到本地挂载信息，和本地的磁盘分区几乎没有差别，只是文件系统的开始是以 IP 地址开头。

NFS 服务所使用的端口号在每次启动的时候是不同的，通过远程过程调用（Remote Procedure Call，RPC）协议/服务来实现，这个 RPC 服务的应用在门户级的网站有很多。如百度。

2. RPC 简介

因为 NFS 支持的功能相当多，而不同的功能都会使用不同的程序来启动，每启动一个功能就会启用一些端口来传输数据，因此，NFS 的功能所对应的端口无法固定，它会随机取用一些未被使用的端口来作为传输之用，其中 CentOS 5.x 的随机端口都小于 1024，而 CentOS 6.x 的随机端口都是较大的。

因为端口不固定，这样就会造成 NFS 客户端与 NFS 服务端的通信障碍，因为 NFS 客户端必须要知道 NFS 服务器端的数据传输端口才能进行通信，才能交互数据。

要解决上面的困扰，就需要通过远程过程调用 RPC 服务来帮忙。NFS 的 RPC 服务最主要的功能就是记录每个 NFS 功能所对应的端口号，并且在 NFS 客户端请求时将该端口和功能对应的信息传递给请求数据的 NFS 客户端，从而确保客户端可以连接到正确的 NFS 端口上去，达到实现数据传输交互数据目的。

3．NFS 相关文件

（1）主要配置文件：/etc/exports。

这是 NFS 的主要配置文件。该文件是空白的，有的系统可能不存在这个文件，主要手动建立。NFS 的配置一般只在这个文件中配置即可。

（2）NFS 文件系统维护指令：/usr/sbin/exportfs。

这是维护 NFS 分享资源的指令，可以利用这个指令重新分享 /etc/exports 变更的目录资源、将 NFS Server 分享的目录卸除或重新分享。

（3）分享资源的登录档：/var/lib/nfs/*tab。

将 NFS 服务器的登录文件都放置到 /var/lib/nfs/ 目录中。在该目录下有两个比较重要的登录档，一个是 etab，主要记录了 NFS 所分享出来的目录的完整权限设定值；另一个 xtab，记录曾经链接到此 NFS 服务器的相关客户端数据。

（4）客户端查询服务器分享资源的指令：/usr/sbin/showmount。

这是另一个重要的 NFS 指令。exportfs 用在 NFS Server 端，而 showmount 主要用在 Client 端。showmount 可以用来察看 NFS 分享出来的目录资源。

5.1.3 任务：部署 NFS 共享存储

任务执行清单：

在本任务中，您将通过实践部署 NFS 共享存储服务器。

目标

- 掌握 NFS 共享服务的安装。
- 掌握 NFS 共享服务的配置。

NFS（网络文件系统）服务可以将远程 Linux 系统上的文件共享资源盖在到本地主机的目录上，从而使得本地主机（Linux 客户端）基于 TCP／IP 协议，像使用本地主机上的资源那样读写远程 Linux 系统上的共享文件。

重要信息

- 本任务采用 CentOS 7 操作系统。
- 虚拟机密码为 000000。
- 提前配置好 yum 源仓库。
- 关闭防火墙和 Selinux。

解决方案

1. 基础环境配置

（1）首先登录到 192.168.200.11 服务器端，更改主机名为 nfs01，使用 yum 软件仓库检查 Linux 系统中是否已经安装了 NFS 软件包。命令和结果如下所示：

```
[root@localhost ~]# hostnamectl set-hostname nfs01
[root@localhost ~]# logout
[root@nfs01 ~]# hostnamectl
   Static hostname: nfs01
         Icon name: computer-vm
           Chassis: vm
        Machine ID: 622ba110a69e24eda2dca57e4d306baa
           Boot ID: 3133e7777c0942fa8a8c2ac486998554
    Virtualization: kvm
  Operating System: CentOS Linux 7 (Core)
       CPE OS Name: cpe:/o:centos:centos:7
            Kernel: Linux 3.10.0-862.2.3.el7.x86_64
      Architecture: x86-64
[root@nfs01 ~]# yum -y install nfs-utils rpcbind
Loaded plugins: fastestmirror
Loading mirror speeds from cached hostfile
centos                                                  | 3.6 kB  00:00:00
mall                                                    | 2.9 kB  00:00:00
Package 1:nfs-utils-1.3.0-0.54.el7.x86_64 already installed and latest version
Package rpcbind-0.2.0-44.el7.x86_64 already installed and latest version
Nothing to do
```

在客户端 192.168.200.12 节点进行同样的操作，更改主机名为 nfs02，并检查 NFS 软件包。

（2）清空 NFS 服务器中 iptables 防火墙的默认策略，以免默认的防火墙策略禁止正常的 NFS 共享服务。

```
[root@nfs01 ~]# iptables -F
[root@nfs01 ~]# iptables-save
```

2. 服务端创建与配置

（1）在 NFS 服务器上建立用于 NFS 文件共享的目录，并设置足够的权限确保他人也有写入权限。

```
[root@nfs01 ~]# mkdir /data
[root@nfs01 ~]# chmod -Rf 777 /data
[root@nfs01 ~]# ll /
drwxrwxrwx   2 root root        6 Jun 22 03:53 data
```

（2）NFS 服务程序配置文件为 /etc/exports，默认情况下里面没有任何内容。可以按照"共享目录的路径 NFS 客户端（共享权限参数）"的格式进行编辑，定义要共享的目录与相应的权限。将表 5-1 中的参数写到 NFS 服务程序的配置文件中。

表 5-1 配置 NFS 服务程序配置文件的参数

参　　数	作　　用
ro	只读
rw	读写
root_squash	当 NFS 客户端以 root 管理员访问时，映射为 NFS 服务器的匿名用户

续表

参数	作用
no_root_squash	当 NFS 客户端以 root 管理员访问时，映射为 NFS 服务器的 root 管理员
all_squash	无论 NFS 客户端使用什么账户访问，均映射为 NFS 服务器的匿名用户
sync	同时将数据写入到内存与硬盘中，保证不丢失数据
async	优先将数据保存到内存，然后再写入硬盘，这样效率更高，但可能会丢失数据

注意，NFS 客户端地址与权限之间没有空格。

```
[root@nfs01 ~]# vi /etc/exports
/data    192.168.200.*(rw,sync,root_squash)
```

（3）启动和启用 NFS 服务程序，在启动 NFS 服务之前，还需要重启并启用 rpcbind 服务程序，并将这两个服务一并加入开机启动项目中。

```
[root@nfs01 ~]# systemctl restart rpcbind
[root@nfs01 ~]# systemctl enable rpcbind
[root@nfs01 ~]# systemctl start nfs-server
[root@nfs01 ~]# systemctl enable nfs-server
Created symlink from /etc/systemd/system/multi-user.target.wants/nfs-server.service to /usr/lib/systemd/system/nfs-server.service.
```

3. 客户端创建与配置

（1）使用 showmount 命令查看 nfs 服务器共享信息。输出格式为"共享的目录名称 允许使用客户端地址"。

```
[root@nfs02 ~]# showmount -e 192.168.200.11
Export list for 192.168.200.11:
/data 192.168.200.*
```

（2）在 NFS 客户端创建一个挂载目录。使用 mount 命令并结合 -t 参数，指定要挂载的文件系统的类型，并在命令后面写上服务器的 IP 地址、服务器上的共享目录以及要挂载到客户端的目录。

```
[root@nfs02 ~]# mkdir /nfs_data
[root@nfs02 ~]# mount -t nfs 192.168.200.11:/data /nfs_data/
```

（3）挂载成功后，在客户端使用 df -h 命令可以看见挂载的详细信息。如果希望 NFS 文件共享服务器能一直有效，则需要将其写入到 fstab 文件中。

```
[root@nfs02 ~]# df -h
Filesystem              Size  Used Avail Use% Mounted on
/dev/vda1                20G  973M   20G   5% /
devtmpfs                474M     0  474M   0% /dev
tmpfs                   496M     0  496M   0% /dev/shm
tmpfs                   496M   13M  483M   3% /run
tmpfs                   496M     0  496M   0% /sys/fs/cgroup
tmpfs                   100M     0  100M   0% /run/user/0
192.168.200.11:/data     20G  1.3G   19G   7% /nfs_data
```

（4）在客户端创建一个文件并添加测试语句，然后登录到服务端查看是否存在，也可以进行一些删除等操作，命令和结果如下所示：

```
[root@nfs02 ~]# cd /nfs_data /
[root@nfs02 nfs_data]# echo "This is client test_file" > client
```

```
[root@nfs01 ~]# cd /data/
[root@nfs01 data]# cat client
This is client test_file
```

本任务到此结束。

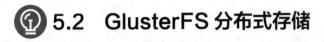 5.2　GlusterFS 分布式存储

学习目标

学完本节后，您应能够：
- 了解什么是 GlusterFS。
- 了解 GlusterFS 工作原理。
- 掌握 GlusterFS 分布式存储服务部署。

5.2.1　GlusterFS 分布式存储概述

　　GlusterFS 是 Scale-Out 存储解决方案 Gluster 的核心，它是一个开源的分布式文件系统，具有强大的横向扩展能力，通过扩展能够支持数 PB 存储容量和处理数千客户端。GlusterFS 借助 TCP/IP 或 InfiniBand RDMA 网络将物理分布的存储资源聚集在一起，使用单一全局命名空间来管理数据。

　　GlusterFS 支持运行在任何标准 IP 网络上标准应用程序的标准客户端，用户可以在全局统一的命令空间中使用 NFS/CIFS 等标准协议来访问应用程序。GlusterFS 使得用户可摆脱原有的独立、高成本的封闭存储系统，能够利用普通廉价的存储设备来部署可集中管理、横向扩展、虚拟化的存储池，存储容量可扩展至 TB/PB 级。

5.2.2　GlusterFS 工作原理

　　1. GlusterFS 堆栈式结构

　　GlusterFS 是根据 fuse 提供的接口实现的一个用户态的文件系统，主要由 gluster、glusterd、glusterfs 和 glusterfsd 四大模块组成。

　　（1）gluster：是 cli 命令执行工具，主要功能是解析命令行参数，然后把命令发送给 glusterd 模块执行。

　　（2）glusterd：是一个管理模块，处理 gluster 发过来的命令，处理集群管理、存储池管理、brick 管理、负载均衡、快照管理等。集群信息、存储池信息和快照信息等都是以配置文件的形式存放在服务器中，当客户端挂载存储时，glusterd 会把存储池的配置文件发送给客户端。

　　（3）glusterfsd：是服务端模块，存储池中的每个 brick 都会启动一个 glusterfsd 进程。此模块主要是处理客户端的读写请求，从关联的 brick 所在磁盘中读写数据，然后返回给客户端。

　　（4）glusterfs：是客户端模块，负责通过 mount 挂载集群中某台服务器的存储池，以目录的形式呈现给用户。当用户从此目录读写数据时，客户端根据从 glusterd 模块获取的存储池的配置文件信息，通过 DHT 算法计算文件所在服务器的 brick 位置，然后通过 Infiniband RDMA 或 TCP/IP 方式把数据发送给 brick，等待 brick 处理完，返回结果给用户。存储池的副本、条带、

hash、EC 等逻辑都在客户端处理。

2. GlusterFS 优点

（1）无元数据节点性能瓶颈。采用无中心对称式架构，没有专用的元数据服务器，也就不存在元数据服务器瓶颈。元数据存在于文件的属性和扩展属性中。

（2）良好的可扩展性。使用弹性 hash 算法代替传统的有元数据节点服务，获得了接近线性的高扩展性。

（3）高可用。采用副本、EC 等冗余设计，保证在冗余范围内的节点掉线时，仍然可以从其他服务节点获取数据，保证高可用性。服务器端还会随着存储池的启动开启一个 glustershd 进程，这个进程会定期检查副本和 EC 卷中各个 brick 之间数据的一致性，并进行恢复。

（4）存储池类型。丰富包括粗粒度、条带、副本、条带副本和 EC，可以根据用户的需求，满足不同程度的冗余。粗粒度卷不带任何冗余，文件不进行切片，是完整地存放在某个 brick 上。

（5）高性能。采用弱一致性的设计，向副本中写数据时，只要有一个 brick 成功返回，就认为写入成功，不必等待其他 brick 返回，这样的方式比强一致性要快。

5.2.3 任务：部署 GlusterFS 分布式存储

在使用 glusterfs 提供的存储服务之前，需要先挂载存储池，向挂载点写数据，会经过 fuse 内核模块传给客户端，客户端检查存储池的类型，然后计算数据所在服务器，最后通过 socket 或 rdma 与服务器通信。

任务执行清单

在本任务中，您将通过实践部署 GlusterFS 分布式存储系统，并使用复制卷和条带卷进行相应实验。

目标

- 掌握 GlusterFS 服务的安装。
- 掌握 GlusterFS 服务的配置。
- 掌握 GlusterFS 复制卷的配置。
- 掌握 GlusterFS 条带卷的配置。

重要信息

- 本任务采用 CentOS 7 操作系统。
- 虚拟机密码为 000000。
- 提前配置好 yum 源仓库。

解决方案

1. 基础环境配置

（1）创建五台虚拟机，节点为 192.168.200.10（20/30/40/50），并且每个节点各添加两块 10 GB 的硬盘为 sdb 和 sdc，此处以 192.168.200.10 节点为例，查看硬盘大小。

```
[root@localhost ~]# lsblk
NAME    MAJ:MIN RM  SIZE RO TYPE MOUNTPOINT
sda       8:0    0   40G  0 disk
├─sda1    8:1    0  500M  0 part /boot
└─sda2    8:2    0 39.5G  0 part
```

```
  ├─centos-root 253:0   0   35.6G  0   lvm  /
  └─centos-swap 253:1   0   3.9G   0   lvm  [SWAP]
sdb              8:16   0   10G    0   disk
sdc              8:32   0   10G    0   disk
sr0              11:0   1   4G     0   rom  /mnt
```

（2）更改五台虚拟机的主机名，第五台为 client 节点，并添加 hosts 文件实现集群之间相互解析，此处以 192.168.200.10 节点为例。

```
[root@localhost ~]# hostnamectl set-hostname glusterfs01
[root@localhost ~]# logout
[root@glusterfs01 ~]# hostnamectl
   Static hostname: glusterfs01
         Icon name: computer-vm
           Chassis: vm
        Machine ID: f6b3e258598b41148bcb855e2c505f94
           Boot ID: 78f09d5ccdb84c9eb2cb909ac0ae5674
    Virtualization: vmware
  Operating System: CentOS Linux 7 (Core)
       CPE OS Name: cpe:/o:centos:centos:7
            Kernel: Linux 3.10.0-327.el7.x86_64
      Architecture: x86-64
[root@glusterfs01 ~]# vi /etc/hosts
127.0.0.1   localhost localhost.localdomain localhost4 localhost4.localdomain4
::1         localhost localhost.localdomain localhost6 localhost6.localdomain6
192.168.200.10 glusterfs01
192.168.200.20 glusterfs02
192.168.200.30 glusterfs03
192.168.200.40 glusterfs04
```

（3）为了不影响后续的连接，五台虚拟机都需要关闭防火墙和 selinux，此处以 192.168.200.10 节点为例，命令和结果如下所示：

```
[root@glusterfs01 ~]# systemctl stop firewalld
[root@glusterfs01 ~]# systemctl disable firewalld
Removed symlink /etc/systemd/system/dbus-org.fedoraproject.FirewallD1.service.
Removed symlink /etc/systemd/system/basic.target.wants/firewalld.service.
[root@glusterfs01 ~]# setenforce 0
```

2. 服务安装与配置

（1）使用 rpm 软件包，充当本地定制化 yum 源文件，此处以 192.168.200.10 节点为例，命令和结果如下所示：

```
[root@glusterfs01 ~]# rm -rf /etc/yum.repos.d/*
[root@glusterfs01 ~]# vi /etc/yum.repos.d/local.repo
[rpm]
name=rpm
baseurl=file:///root/rpm
gpgcheck=0
enabled=1
[centos]
name=centos
baseurl=file:///mnt
gpgcheck=0
enabled=1
```

（2）安装 GlusterFS 分布式存储服务，并查看版本信息，此处以 192.168.200.10 节点为例，命令和结果如下所示：

```
[root@glusterfs01 ~]# yum -y install glusterfs-server glusterfs-cli glusterfs-geo-replication
[root@glusterfs01 ~]# which glusterfs
/usr/sbin/glusterfs
[root@glusterfs01 ~]# glusterfs -V
glusterfs 3.7.20 built on Jan 30 2017 15:39:27
Repository revision: git://git.gluster.com/glusterfs.git
Copyright (c) 2006-2013 Red Hat, Inc. <http://www.redhat.com/>
GlusterFS comes with ABSOLUTELY NO WARRANTY.
It is licensed to you under your choice of the GNU Lesser
General Public License, version 3 or any later version (LGPLv3
or later), or the GNU General Public License, version 2 (GPLv2),
in all cases as published by the Free Software Foundation.
```

（3）启动 GlusterFS 服务并设为开机自启，启动成功后查看服务状态信息，此处以 192.168.200.10 节点为例，命令和结果如下所示：

```
[root@glusterfs01 ~]# systemctl start glusterd
[root@glusterfs01 ~]# chkconfig glusterd on
[root@glusterfs01 ~]# systemctl status glusterd
[root@glusterfs01 ~]# systemctl status glusterd
● glusterd.service - LSB: glusterfs server
   Loaded: loaded (/etc/rc.d/init.d/glusterd)
   Active: active (running) since Sun 2021-04-11 18:59:52 EDT; 1min 48s ago
     Docs: man:systemd-sysv-generator(8)
   CGroup: /system.slice/glusterd.service
           └─10549 /usr/sbin/glusterd --pid-file=/run/glusterd.pid
Apr 11 18:59:50 glusterfs01 systemd[1]: Starting LSB: glusterfs server...
Apr 11 18:59:52 glusterfs01 glusterd[10542]: Starting glusterd:[  OK  ]
Apr 11 18:59:52 glusterfs01 systemd[1]: Started LSB: glusterfs server.
```

（4）存储主机加入信任存储池，此处只需要在一台虚拟机上进行添加操作即可，但并不需要添加信任自己。确保所有的虚拟机的 glusterd 服务都处于开启状态，然后执行如下操作：

```
[root@glusterfs01 ~]# gluster peer probe glusterfs02
peer probe: success.
[root@glusterfs01 ~]# gluster peer probe glusterfs03
peer probe: success.
[root@glusterfs01 ~]# gluster peer probe glusterfs04
peer probe: success.
```

（5）登录到不同的主机，查看虚拟机信任状态结果是否添加成功，命令和结果如下所示：

```
[root@glusterfs01 ~]# gluster peer status
Number of Peers: 3
Hostname: glusterfs02
Uuid: 0f2c38ec-902d-4c03-b93b-0f03cc803d94
State: Peer in Cluster (Connected)
Hostname: glusterfs03
Uuid: 6e808333-5cf2-4ee0-80e5-0b727e786b29
State: Peer in Cluster (Connected)
Hostname: glusterfs04
Uuid: 1b0a39f4-36dc-4c88-9ed6-a93402c7de8f
```

```
State: Peer in Cluster (Connected)
[root@glusterfs02 ~]# gluster peer status
Number of Peers: 3
Hostname: glusterfs01
Uuid: df53b669-73aa-40b7-96d8-19aeaa3f0082
State: Peer in Cluster (Connected)
Hostname: glusterfs03
Uuid: 6e808333-5cf2-4ee0-80e5-0b727e786b29
State: Peer in Cluster (Connected)
Hostname: glusterfs04
Uuid: 1b0a39f4-36dc-4c88-9ed6-a93402c7de8f
State: Peer in Cluster (Connected)
```

（6）安装 xfsprogs 格式化工具，格式化每台虚拟机的那两块 10 GB 硬盘，然后进行挂载查看，此处以 192.168.200.10 节点为例。

```
[root@glusterfs01 ~]# yum -y install xfsprogs
[root@glusterfs01 ~]# ll /dev/sd*
brw-rw----. 1 root disk 8,  0 Apr 11 17:15 /dev/sda
brw-rw----. 1 root disk 8,  1 Apr 11 17:15 /dev/sda1
brw-rw----. 1 root disk 8,  2 Apr 11 17:15 /dev/sda2
brw-rw----. 1 root disk 8, 16 Apr 11 17:15 /dev/sdb
brw-rw----. 1 root disk 8, 32 Apr 11 17:15 /dev/sdc
[root@glusterfs01 ~]# mkfs.ext4 /dev/sdb
mke2fs 1.42.9 (28-Dec-2013)
/dev/sdb is entire device, not just one partition!
Proceed anyway? (y,n) y
Filesystem label=
OS type: Linux
Block size=4096 (log=2)
Fragment size=4096 (log=2)
Stride=0 blocks, Stripe width=0 blocks
655360 inodes, 2621440 blocks
131072 blocks (5.00%) reserved for the super user
First data block=0
Maximum filesystem blocks=2151677952
80 block groups
32768 blocks per group, 32768 fragments per group
8192 inodes per group
Superblock backups stored on blocks:
        32768, 98304, 163840, 229376, 294912, 819200, 884736, 1605632
Allocating group tables: done
Writing inode tables: done
Creating journal (32768 blocks): done
Writing superblocks and filesystem accounting information: done
[root@glusterfs01 ~]# mkdir -p /gluster/brick1
[root@glusterfs01 ~]# mount /dev/sdb /gluster/brick1/
[root@glusterfs01 ~]# df -h
Filesystem               Size  Used Avail Use% Mounted on
/dev/mapper/centos-root   36G  1.7G   34G   5% /
devtmpfs                 479M     0  479M   0% /dev
tmpfs                    489M     0  489M   0% /dev/shm
tmpfs                    489M  6.9M  483M   2% /run
tmpfs                    489M     0  489M   0% /sys/fs/cgroup
```

```
/dev/sda1                  497M  114M  384M  23% /boot
tmpfs                       98M    0    98M   0% /run/user/0
/dev/sr0                   4.1G  4.1G     0 100% /mnt
/dev/sdb                   9.8G   37M  9.2G   1% /gluster/brick1
[root@glusterfs01 ~]# mkfs.ext4 /dev/sdc
mke2fs 1.42.9 (28-Dec-2013)
/dev/sdc is entire device, not just one partition!
Proceed anyway? (y,n) y
Filesystem label=
OS type: Linux
Block size=4096 (log=2)
Fragment size=4096 (log=2)
Stride=0 blocks, Stripe width=0 blocks
655360 inodes, 2621440 blocks
131072 blocks (5.00%) reserved for the super user
First data block=0
Maximum filesystem blocks=2151677952
80 block groups
32768 blocks per group, 32768 fragments per group
8192 inodes per group
Superblock backups stored on blocks:
        32768, 98304, 163840, 229376, 294912, 819200, 884736, 1605632
Allocating group tables: done
Writing inode tables: done
Creating journal (32768 blocks): done
Writing superblocks and filesystem accounting information: done
[root@glusterfs01 ~]# mkdir -p /gluster/brick2
[root@glusterfs01 ~]# mount /dev/sdc /gluster/brick2
[root@glusterfs01 ~]# df -h
Filesystem                 Size  Used Avail Use% Mounted on
/dev/mapper/centos-root     36G  1.7G   34G   5% /
devtmpfs                   479M     0  479M   0% /dev
tmpfs                      489M     0  489M   0% /dev/shm
tmpfs                      489M  6.9M  483M   2% /run
tmpfs                      489M     0  489M   0% /sys/fs/cgroup
/dev/sda1                  497M  114M  384M  23% /boot
tmpfs                       98M     0   98M   0% /run/user/0
/dev/sr0                   4.1G  4.1G     0 100% /mnt
/dev/sdb                   9.8G   37M  9.2G   1% /gluster/brick1
/dev/sdc                   9.8G   37M  9.2G   1% /gluster/brick2
```

3. 分布式卷部署

（1）创建 volume 分布式卷，在 glusterfs01 节点创建成功之后启动创建的卷，命令和结果如下所示：

```
[root@glusterfs01 ~]# gluster volume create gs1 glusterfs01:/gluster/brick1 glusterfs02:/gluster/brick1 force
volume create: gs1: success: please start the volume to access data
[root@glusterfs01 ~]# gluster volume start gs1
volume start: gs1: success
```

（2）登录到不同的虚拟机可以看到 volume 卷的详细信息，命令和结果如下所示：

```
[root@glusterfs01 ~]# gluster volume info
Volume Name: gs1
```

```
Type: Distribute
Volume ID: 58c45b1d-3c80-41a1-aef7-6f8a464b3aef
Status: Started
Number of Bricks: 2
Transport-type: tcp
Bricks:
Brick1: glusterfs01:/gluster/brick1
Brick2: glusterfs02:/gluster/brick1
[root@glusterfs03 ~]# gluster volume info
Volume Name: gs1
Type: Distribute
Volume ID: 58c45b1d-3c80-41a1-aef7-6f8a464b3aef
Status: Started
Number of Bricks: 2
Transport-type: tcp
Bricks:
Brick1: glusterfs01:/gluster/brick1
Brick2: glusterfs02:/gluster/brick1
Options Reconfigured:
performance.readdir-ahead: on
```

（3）使用 glusterfs 方式挂载 volume 卷，此处也可以使用 nfs 挂载，将服务挂载到指定目录之后，在挂载好的 /opt 目录里创建一个实验文件，命令和结果如下所示：

```
[root@glusterfs01 ~]# mount -t glusterfs 127.0.0.1:/gs1 /opt
[root@glusterfs01 ~]# df -h
Filesystem              Size  Used Avail Use% Mounted on
/dev/mapper/centos-root  36G  1.7G   34G   5% /
devtmpfs                479M     0  479M   0% /dev
tmpfs                   489M     0  489M   0% /dev/shm
tmpfs                   489M  6.9M  483M   2% /run
tmpfs                   489M     0  489M   0% /sys/fs/cgroup
/dev/sda1               497M  114M  384M  23% /boot
tmpfs                    98M     0   98M   0% /run/user/0
/dev/sr0                4.1G  4.1G     0 100% /mnt
/dev/sdb                9.8G   37M  9.2G   1% /gluster/brick1
/dev/sdc                9.8G   37M  9.2G   1% /gluster/brick2
127.0.0.1:/gs1           20G   73M   19G   1% /opt
[root@glusterfs01 ~]# touch /opt/{1..5}
[root@glusterfs01 ~]# ls /opt/
1  2  3  4  5  lost+found
```

（4）登录到其他虚拟机，挂载分布式卷 gs1 到对应目录下，使用命令查看同步挂载结果。此处以 glusterfs02 和 glusterfs04 为例。

```
[root@glusterfs02 ~]# mount -t glusterfs 127.0.0.1:/gs1 /opt
[root@glusterfs02 ~]# ls /opt/
1  2  3  4  5  lost+found
[root@glusterfs04 ~]# mount -t glusterfs 127.0.0.1:/gs1 /opt
[root@glusterfs04 ~]# ls /opt/
1  2  3  4  5  lost+found
```

（5）在 glusterfs01 和 glusterfs02 节点中查询实验文件结果，发现两个节点存储的信息数据是不相同的，说明分布式存储的实验部署成功。命令和结果如下所示：

```
[root@glusterfs01 ~]# ls /gluster/brick1
1  5  lost+found
[root@glusterfs02 ~]# ls /gluster/brick1
2  3  4  lost+found
```

4. 创建分布式复制卷和条带卷

（1）在任意一台 gluster 虚拟机上，创建 volume 分布式卷 gs2，在 glusterfs02 节点创建成功之后启动创建的卷，并查看卷信息。命令和结果如下所示：

```
[root@glusterfs02 ~]# gluster volume create gs2 replica 2 glusterfs03:/gluster/brick1 glusterfs04:/gluster/brick1 force
volume create: gs2: success: please start the volume to access data
[root@glusterfs02 ~]# gluster volume start gs2
volume start: gs2: success
[root@glusterfs02 ~]# gluster volume info gs2
Volume Name: gs2
Type: Replicate                              #复制卷
Volume ID: 698d45c7-57c4-40e8-85e3-ebe9eb426abb
Status: Started
Number of Bricks: 1 × 2 = 2
Transport-type: tcp
Bricks:
Brick1: glusterfs03:/gluster/brick1
Brick2: glusterfs04:/gluster/brick1
Options Reconfigured:
performance.readdir-ahead: on
```

（2）在任意一台 gluster 虚拟机上，创建 stripe 条带卷 gs3，在 glusterfs02 节点创建成功之后启动创建的卷，并查看卷信息，命令和结果如下所示：

```
[root@glusterfs02 ~]# gluster volume create gs3 stripe 2 glusterfs01:/gluster/brick2 glusterfs02:/gluster/brick2 force
volume create: gs3: success: please start the volume to access data
[root@glusterfs02 ~]# gluster volume start gs3
volume start: gs3: success
[root@glusterfs02 ~]# gluster volume info gs3
Volume Name: gs3
Type: Stripe                                 #条带卷
Volume ID: e3d20271-efae-41b8-b761-da21e37694a1
Status: Started
Number of Bricks: 1 × 2 = 2
Transport-type: tcp
Bricks:
Brick1: glusterfs01:/gluster/brick2
Brick2: glusterfs02:/gluster/brick2
Options Reconfigured:
performance.readdir-ahead: on
```

（3）在 glusterfs01 节点，安装 nfs 服务器，启动之后查看卷服务状态。命令和结果如下所示：

```
[root@glusterfs01 ~]# yum -y install nfs-utils
[root@glusterfs01 ~]# systemctl start rpcbind
[root@glusterfs01 ~]# systemctl start nfs-utils
[root@glusterfs01 ~]# systemctl stop glusterd
```

```
[root@glusterfs01 ~]# systemctl start glusterd
[root@glusterfs01 ~]# gluster volume status
Status of volume: gs1
Gluster process                              TCP Port  RDMA Port  Online  Pid
------------------------------------------------------------------------------
Brick glusterfs01:/gluster/brick1            49152     0          Y       10918
Brick glusterfs02:/gluster/brick1            49152     0          Y       10871
NFS Server on localhost                      2049      0          Y       11650
NFS Server on glusterfs03                    N/A       N/A        N       N/A
NFS Server on glusterfs02                    N/A       N/A        N       N/A
NFS Server on glusterfs04                    N/A       N/A        N       N/A
Task Status of Volume gs1
------------------------------------------------------------------------------
There are no active volume tasks
```

5. 客户端配置与测试

（1）首先连接客户端192.168.200.50，更改主机名之后，使用yum安装nfs服务器并启动，命令如下所示：

```
[root@localhost ~]# hostnamectl set-hostname client
[root@localhost ~]# logout
[root@localhost ~]# yum -y install nfs-utils
[root@client ~]# systemctl start rpcbind
[root@client ~]# systemctl start nfs-utils
```

（2）分布式卷gs1的数据写入测试。在客户端创建一个测试目录将gs1进行nfs卷挂载操作，在client上进行数据读写操作，然后登录glusterfs01和glusterfs02进行数据的查看，命令和结果如下所示：

```
[root@client ~]# mkdir /opt/gs1
[root@client ~]# mount -t nfs 192.168.200.10:/gs1 /opt/gs1/
[root@client ~]# touch /opt/gs1/{1..10}
[root@client ~]# ls /opt/gs1/
1  10  2  3  4  5  6  7  8  9  lost+found
[root@glusterfs01 ~]# ls /gluster/brick1
1  5  7  8  9  lost+found
[root@glusterfs02 ~]# ls /gluster/brick1/
10  2  3  4  6  lost+found
```

分布式卷的数据存储方式是将数据平均写入到每个整合的磁盘中，类似于raid0，写入速度快，但这样磁盘一旦损坏没有纠错能力。

（3）分布式复制卷gs2的数据写入测试。在客户端创建一个测试目录，将gs2进行nfs卷挂载操作，在client上进行数据读写操作，然后登录glusterfs03和glusterfs04进行数据查看。命令和结果如下所示：

```
[root@client ~]# mkdir /opt/gs2
[root@client ~]# mount -t nfs 192.168.200.10:/gs2 /opt/gs2/
[root@client ~]# ls /opt/gs2/
lost+found
[root@client ~]# ls /opt/gs2/
1  10  2  3  4  5  6  7  8  9  lost+found
[root@glusterfs03 ~]# ls /gluster/brick1/
1  10  2  3  4  5  6  7  8  9  lost+found
```

```
[root@glusterfs04 ~]# ls /gluster/brick1/
1 10 2 3 4 5 6 7 8 9 lost+found
```

分布式复制卷的数据存储方式为：每个整合的磁盘中都写入同样的数据内容，类似于 raid1，数据非常安全、读取性能高、占磁盘容量高。

（4）分布式条带卷 gs3 的数据写入测试。在客户端创建一个测试目录将 gs3 进行 nfs 卷挂载操作，在 client 上进行数据读写操作，然后登录 glusterfs01 和 glusterfs02 进行数据的查看，命令和结果如下所示：

```
[root@client ~]# mkdir /opt/gs3
[root@client ~]# mount -o nolock -t nfs 192.168.200.10:/gs3 /opt/gs3/
[root@client ~]# dd if=/dev/zero of=/root/test bs=1024 count=262144
262144+0 records in
262144+0 records out
268435456 bytes (268 MB) copied, 5.77319 s, 46.5 MB/s
[root@client ~]# ls
anaconda-ks.cfg  test
[root@client ~]# cp test /opt/gs3/
[root@client ~]# ls /opt/gs3/
lost+found  test
[root@client ~]# du -sh /opt/gs3/test
256M    /opt/gs3/test
[root@glusterfs01 ~]# du -sh /gluster/brick2/test
129M    /gluster/brick2/test
[root@glusterfs02 ~]# du -sh /gluster/brick2/test
129M    /gluster/brick2/test
```

分布式条带卷是将数据的容量平均分配到了每个整合的磁盘节点上，可大幅提高大文件的并发读访问。

本任务到此结束。

 5.3 Ceph 分布式存储

学习目标

学完本节后，您应能够：
- 了解什么是 Ceph。
- 了解 Ceph 工作原理。
- 掌握 Ceph 分布式存储服务部署。

5.3.1 Ceph 分布式存储概述

1．Ceph 简介

随着 OpenStack 日渐成为开源云计算的标准软件栈，Ceph 也已经成为 OpenStack 的首选后端存储。Ceph 是一种为优秀的性能、可靠性和可扩展性而设计的统一的、分布式文件系统。

Ceph 是一个开源的分布式文件系统。因为它还支持块存储、对象存储，所以很自然地被用作云计算框架 OpenStack 或 CloudStack 的整个存储后端。当然也可以单独作为存储，例如部署

一套集群作为对象存储、SAN 存储、NAS 存储等。

2. Ceph 存储的优点

(1) 统一存储。虽然 Ceph 底层是一个分布式文件系统，但由于在上层开发了支持对象和块的接口。所以在开源存储软件中，能够得到广泛应用。

(2) 高扩展性。扩容方便、容量大。能够管理上千台服务器、EB 级的容量。

(3) 可靠性强。支持多份强一致性副本，EC。副本能够跨主机、机架、机房、数据中心存放。所以安全可靠。存储节点可以自动管理、自动修复、无单点故障、容错性强。

(4) 高性能。因为是多个副本，因此在读写操作时能够做到高度并行化。理论上，节点越多整个集群的 IOPS 和吞吐量越高。另外，ceph 客户端读写数据可直接与存储设备（osd）交互。

3. Ceph 组件介绍

Ceph OSDs: Ceph OSD 守护进程（Ceph OSD）的功能是存储数据，处理数据的复制、恢复、回填、再均衡，并通过检查其他 OSD 守护进程的心跳来向 Ceph Monitors 提供一些监控信息。当 Ceph 存储集群设定为有 2 个副本时，至少需要 2 个 OSD 守护进程，集群才能达到 active+clean 状态（Ceph 默认有 3 个副本，但用户可以调整副本数）。

Monitors: Ceph Monitor 维护着展示集群状态的各种图表，包括监视器图、OSD 图、归置组（PG）图和 CRUSH 图。Ceph 保存着发生在 Monitors、OSD 和 PG 上的每一次状态变更的历史信息（称为 epoch）。

MDSs: Ceph 元数据服务器（MDS）为 Ceph 文件系统存储元数据（也就是说，Ceph 块设备和 Ceph 对象存储不使用 MDS）。元数据服务器使得 POSIX 文件系统的用户可以在不对 Ceph 存储集群造成负担的前提下，执行 ls、find 等基本命令。

5.3.2 Ceph 工作原理

1. Ceph 服务原理

Ceph 作为 Linux PB 级分布式文件系统，因其灵活、智能、可配置，在软件定义存储的大潮中，越来越受到 IaaS 方案提供商的注意。

OpenStack 中围绕虚拟机主要的存储需求来自 Nova 中的 Disk、Glance 中的 Image、Cinder 中的虚拟硬盘，Ceph 架构如图 5-1 所示。

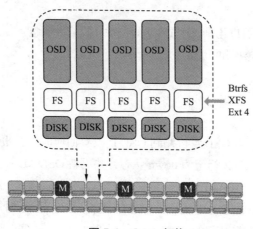

图 5-1　Ceph 架构

2. RADOS 组成

Ceph 的底层是 RADOS，它的意思是 A reliable, autonomous, distributed object storage（一个可靠的、自主的、分布式对象存储系统）。RADOS 由两个组件组成：

（1）OSD：Object Storage Device，提供存储资源。

（2）Monitor：维护整个 Ceph 集群的全局状态。

RADOS 具有很强的可扩展性和可编程性，Ceph 基于 RADOS 开发了 Object Storage（对象存储）、Block Storage（块存储）和 File System（文件系统）。

3. Ceph 映射

Ceph 的命名空间是（Pool,Object），每个 Object 都会映射到一组 OSD 中，由这组 OSD 保存这个 Object，其映射关系如下：

$$(Pool,Object) \rightarrow (Pool,PG) \rightarrow OSD\ SET \rightarrow Disk$$

Ceph 中 Pools 的属性包括：

（1）Object 的副本数。

（2）Placement Groups（数据映射）的数量。

（3）所使用的 CRUSH Ruleset。

在 Ceph 中，Object 先映射到 PG（Placement Group），再由 PG 映射到 OSD SET。每个 Pool 有多个 PG，每个 Object 通过计算 Hash 值并得到它所对应的 PG。PG 再映射到一组 OSD（OSD 的个数由 Pool 的副本数决定），第一个 OSD 是 Primary，剩下的都是 Replicas。

数据映射的方式决定了存储系统的性能和扩展性。（Pool，PG）→ OSD SET 的映射由 4 个因素决定：

（1）CRUSH 算法：一种伪随机算法。

（2）OSD MAP：包含当前所有 Pool 的状态和所有 OSD 的状态。

（3）CRUSH MAP：包含当前磁盘、服务器、机架的层级结构。

（4）CRUSH Rules：数据映射的策略。这些策略可以灵活地设置 Object 存放的区域。

4. 数据映射的优点

（1）把 Object 分组，这降低了需要追踪和处理的 Metadata（元数据）数量（在全局的层面上，不需要追踪和处理每个 Object 的 Metadata 和 Placement，只需要管理 PG 的 Metadata 就可以了。PG 的数量级远远低于 Object 的数量级）。

（2）增加 PG 的数量，可以以均衡每个 OSD 的负载，提高并行度。

（3）分隔故障域，提高数据的可靠性。

（4）强一致性。

（5）容错性。

5. 故障检测

OSD 之间有心跳检测，当 OSD A 检测到 OSD B 没有回应时，会报告给 Monitors 说 OSD B 无法连接，则 Monitors 给 OSD B 标记为 down 状态，并更新 OSD Map。

6. 故障恢复

当某个 PG 对应的 OSD SET 中有一个 OSD 被标记为 down 时（如果 Primary 被标记为 down，则某个 Replica 会成为新的 Primary，并处理所有读写 Object 请求），则该 PG 处于 active+degraded 状态，也就是当前 PG 有效的副本数是 $N-1$。

当 Peering 过程完成之后，PG 进入 active+recoverying 状态，Primary 会迁移和同步那些降级的 Objects 到所有的 Replicas 上，保证这些 Objects 的副本数为 N。

5.3.3 任务：安装 Ceph 服务

目前 Ceph 有很多用户案例，从调查中可以看到有 26% 的用户在生产环境中使用 Ceph，有 37% 的用户在私有云中使用 Ceph，还有 16% 的用户在公有云中使用 Ceph。

任务执行清单

在本任务中，您将通过实践进行 Ceph 分布式存储服务的安装。

目标

- 掌握 Ceph 分布式存储系统的安装。
- 掌握网络 yum 源文件的配置。

重要信息

- 本任务采用 CentOS 7 操作系统。
- 虚拟机密码为 000000。
- 关闭防火墙和 Selinux。

解决方案

1. 基础环境配置

（1）创建 3 个 CentOS 7 系统虚拟机，并修改 hostname 为 ceph-node1、ceph-node2 和 ceph-node3。此处以 ceph-node1 节点为例。

```
[root@ceph-nod1 ~]# hostnamectl set-hostname ceph-node1
[root@ceph-nod1 ~]# logout
[root@ceph-node1 ~]#
```

（2）更改三台虚拟机的主机名，并添加 hosts 文件实现 ceph 服务之间相互解析，此处以 ceph-node1 节点为例。

```
[root@ceph-node1 ~]# vi /etc/hosts
127.0.0.1      localhost localhost.localdomain localhost4 localhost4.localdomain4
::1            localhost localhost.localdomain localhost6 localhost6.localdomain6
192.168.200.10 ceph-node1
192.168.200.20 ceph-node2
192.168.200.30 ceph-node3
```

（3）设置免密码登录，在管理节点使用 ssh-keygen 生成 ssh keys 发布到各节点。

```
[root@ceph-node1 ~]# ssh-keygen
[root@ceph-node1 ~]# ssh-copy-id ceph-node1
[root@ceph-node1 ~]# ssh-copy-id ceph-node2
[root@ceph-node1 ~]# ssh-copy-id ceph-node3
```

（4）安装 NTP 服务，使用互联网上提供的 NTP 服务，让 3 台主机时间保持一致（主机需要可以访问互联网）。

```
[root@ceph-node1 ~]# yum -y install ntp
[root@ceph-node1 ~]# ntpdate ntp1.aliyun.com
[root@ceph-node2 ~]# ntpdate ntp1.aliyun.com
[root@ceph-node3 ~]# ntpdate ntp1.aliyun.com
```

（5）各节点增加 yum 配置文件，首先将原来的 yum 文件移走，然后下载网络 yum 源，此处以 ceph-node1 节点为例。

```
[root@ceph-node1 ~]# mkdir /etc/yum.repos.d/yum/
[root@ceph-node1 ~]# mv /etc/yum.repos.d/*.repo /etc/yum.repos.d/yum/
[root@ceph-node1 ~]# wget -O /etc/yum.repos.d/CentOS-Base.repo http://mirrors.aliyun.com/repo/Centos-7.repo
[root@ceph-node1 ~]# wget -O /etc/yum.repos.d/epel.repo http://mirrors.aliyun.com/repo/epel-7.repo
[root@ceph-node1 ~]# vi /etc/yum.repos.d/ceph.repo
[ceph]
name=ceph
baseurl=http://mirrors.aliyun.com/ceph/rpm-jewel/el7/x86_64/
gpgcheck=0
priority=1
[ceph-noarch]
name=cephnoarch
baseurl=http://mirrors.aliyun.com/ceph/rpm-jewel/el7/noarch/
gpgcheck=0
priority=1
[ceph-source]
name=Ceph source packages
baseurl=http://mirrors.aliyun.com/ceph/rpm-jewel/el7/SRPMS/
gpgcheck=0
priority=1
[root@ceph-node1 ~]# yum clean all
[root@ceph-node1 ~]# yum makecache
```

2. 安装 ceph 服务

（1）在配置好软件源文件之后，使用命令安装 ceph-deploy 管理工具，此时会在 /etc/ceph 路径下生成三个文件，命令和结果如下所示：

```
[root@ceph-node1 ~]# yum -y install ceph-deploy
[root@ceph-node1 ~]# cd /etc/ceph/
[root@ceph-node1 ceph]# ls
ceph.conf  ceph-deploy-ceph.log  ceph.mon.keyring
```

（2）在 ceph-node1 节点上安装 ceph 二进制软件包，使用 ceph-deploy 工具在所有节点上安装。

```
[root@ceph-node1 ceph]#  ceph-deploy install ceph-node1 ceph-node2 ceph-node3
```

如果网络源安装失败，先手工安装 epel-release，然后安装 yum -y install ceph-release，最后安装 yum -y install ceph ceph-radosgw。

（3）登录到虚拟机节点查看 ceph 服务版本信息，命令和结果如下所示：

```
[root@ceph-node1 ceph]# ceph -v
ceph version 10.2.11 (e4b061b47f07f583c92a050d9e84b1813a35671e)
[root@ceph-node2 ~]# ceph -v
ceph version 10.2.11 (e4b061b47f07f583c92a050d9e84b1813a35671e)
[root@ceph-node3 ~]# ceph -v
ceph version 10.2.11 (e4b061b47f07f583c92a050d9e84b1813a35671e)
```

本任务到此结束。

5.4 Swift 分布式存储

学习目标

学完本节后,您应能够:
- 了解什么是 Swift。
- 了解 Swift 工作原理。
- 掌握 Swift 分布式存储服务部署。

5.4.1 Swift 分布式存储概述

1. Swift 简介

在 IT 运维的各种架构中,分布式存储对架构的整体 IO 性能影响至关重要,下面给大家介绍一款开源的分布式存储 Swift。最初 Swift 是提供高可用分布式对象存储的服务,为 nova 组件提供虚拟机镜像存储服务。在数据冗余方面,无须采用 read 在软件层面引入一致性散列技术和数据冗余,牺牲一定程度的数据一致性,来达到高可用和可伸缩性。支持多租户模式下,容器和对象读写操作,适用于互联网应用场景下非结构化的数据存储。

OpenStack Swift 开源项目提供了弹性可伸缩、高可用的分布式对象存储服务,适合存储大规模非结构化数据。

Swift 并不是文件系统或者实时的数据存储系统,它是对象存储,用于永久类型的静态数据的长期存储,这些数据可以检索、调整,必要时进行更新。最适合存储的数据类型是虚拟机镜像、图片、邮件和存档备份。

2. Swift 基本原理

(1)一致性散列(Consistent Hashing)。

面对海量级别的对象,需要存放在成千上万台服务器和硬盘设备上,首先要解决寻址问题,即如何将对象分布到这些设备地址上。Swift 是基于一致性散列技术,通过计算可将对象均匀分布到虚拟空间的虚拟节点上,在增加或删除节点时可大大减少需移动的数据量。

(2)数据一致性模型(Consistency Model)。

按照 Eric Brewer 的 CAP(Consistency、Availability、Partition Tolerance)理论,无法同时满足 3 个方面,Swift 放弃严格一致性(满足 ACID 事务级别),而采用最终一致性模型(Eventual Consistency),来达到高可用性和无限水平扩展能力。

(3)环的数据结构。

环是为了将虚拟节点(分区)映射到一组物理存储设备上,并提供一定的冗余度而设计的,如图 5-2 所示,以查找一个对象的计算过程为例。

(4)数据模型。

Swift 采用层次数据模型,共设三层逻辑结构:Account/Container/Object(即账户/容器/对象),每层节点数均没有限制,可以任意扩展。这里的账户和个人账户不是一个概念,可理解为租户,用来做顶层的隔离机制,可以被多个个人账户共同使用;容器代表封装一组对象,类似文件夹或目录;叶子节点代表对象,由元数据和内容两部分组成,如图 5-3 所示。

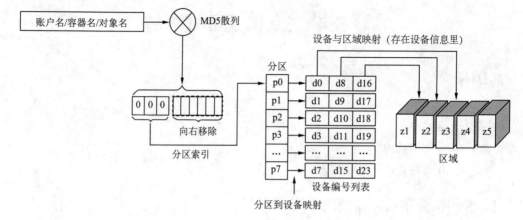

图 5-2 环的数据架构

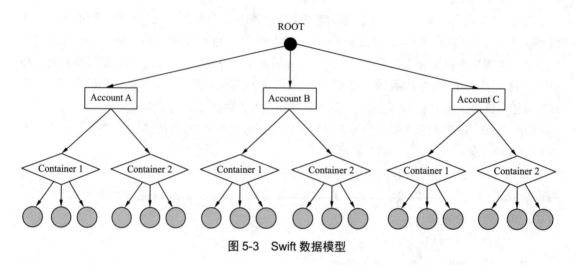

图 5-3 Swift 数据模型

5.4.2 Swift 服务架构

1. Swift 服务架构

Swift 采用完全对称、面向资源的分布式系统架构设计，所有组件都可扩展，避免因单点失效而扩散并影响整个系统运转。通信方式采用非阻塞式 I/O 模式，提高了系统吞吐和响应能力。图 5-4 所示是 Swift 系统架构图。

2. Swift 服务组件介绍

（1）代理服务（Proxy Server）：对外提供对象服务 API，会根据环的信息来查找服务地址并转发用户请求至相应的账户、容器或者对象服务。

（2）认证服务（Authentication Server）：验证访问用户的身份信息，并获得一个对象访问令牌（Token），在一定的时间内会一直有效。

（3）缓存服务（Cache Server）：缓存的内容包括对象服务令牌，账户和容器的存在信息，但不会缓存对象本身的数据。

（4）账户服务（Account Server）：提供账户元数据和统计信息，并维护所含容器列表的服务，每个账户的信息被存储在一个 SQLite 数据库中。

（5）容器服务（Container Server）：提供容器元数据和统计信息，并维护所含对象列表的服务，每个容器的信息也存储在一个 SQLite 数据库中。

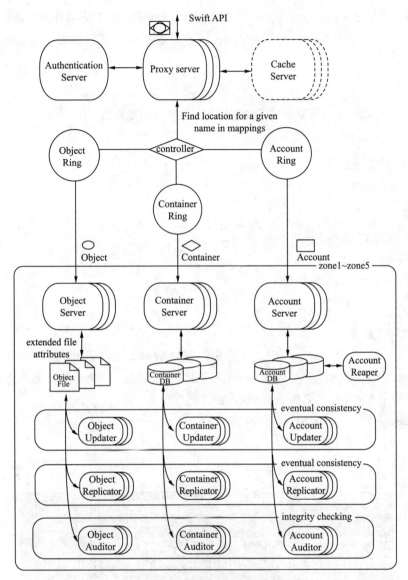

图 5-4　Swift 系统架构图

（6）对象服务（Object Server）：提供对象元数据和内容服务，每个对象的内容会以文件的形式存储在文件系统中，元数据会作为文件属性来存储。

（7）复制服务（Replicator）：会检测本地分区副本和远程副本是否一致，具体是通过对比散列文件和高级水印来完成，发现不一致时会采用推式（Push）更新远程副本。

（8）更新服务（Updater）：当对象由于高负载的原因而无法立即更新时，任务将会被序列化到在本地文件系统中进行排队，以便服务恢复后进行异步更新。

（9）审计服务（Auditor）：检查对象，容器和账户的完整性，如果发现比特级的错误，文件将被隔离，并复制其他的副本以覆盖本地损坏的副本。

（10）账户清理服务（Account Reaper）：移除被标记为删除的账户，删除其所包含的所有容器和对象。

5.4.3 任务：使用 Swift 作为后端存储

Swift 服务是属于 OpenStack 中的一种组件服务。OpenStack 中的组件服务还有 Keystone、Nova、Glance 等，不同的服务负责不同的功能，下面是用 Swift 组件服务作为 Glance 镜像服务后端存储的实验。

任务执行清单

在本任务中，您将通过对象存储 Swift 作为 Glance 镜像服务的后端存储。

目标

- 掌握 Swift 服务配置。
- 掌握 Swift 作为 Glance 后端存储。

重要信息

- 本任务采用 CentOS 7 操作系统。
- 虚拟机密码为 000000。
- 提前搭建好 OpenStack 私有云。

解决方案

1. 基础环境配置

（1）默认情况下，OpenStack Glance 将映像和 OpenStack 实例快照保存到 /var/lib/glance/images/ 的本地文件系统中。首先以 root 用户身份登录到运行 Glance 服务的节点，然后获取 OpenStack 凭据（通常是一个名为 openrc.sh 的文件）。

```
[root@controller ~]# source /etc/keystone/admin-openrc.sh
```

（2）查看 admin 用户的角色信息，命令如下所示，结果如图 5-5 所示。

```
[root@controller ~]# openstack role list
```

图 5-5 角色信息

（3）由于提前安装好了 Swift 服务，此处我们查看 Swift 服务的状态信息。命令和结果如下所示：

```
[root@controller ~]# swift stat
                        Account: AUTH_263606946e8d43d28d094cbea748b9f2
                     Containers: 1
                        Objects: 4
                          Bytes: 510459904
  Containers in policy "policy-0": 1
     Objects in policy "policy-0": 4
       Bytes in policy "policy-0": 510459904
      X-Account-Project-Domain-Id: bda911602b22428fab690543ead44c9f
          X-Openstack-Request-Id: txe37f822d22b24f5eb39b7-006076e550
```

```
         X-Timestamp: 1618398821.42229
          X-Trans-Id: txe37f822d22b24f5eb39b7-006076e550
        Content-Type: application/json; charset=utf-8
        Accept-Ranges: bytes
```

（4）使用命令查看容器的列表，显示为空。

```
[root@controller ~]# swift list
```

2. 配置 Swift 作为 Glance 后端存储

（1）登录到 OpenStack 控制节点，由于提前安装好了各个组件，所以此处直接进入文件配置，更改 Swift 与 Glance 后端配置信息，打开 /etc/glance/glance-swift.conf 更改信息，随后重启服务并查看镜像列表。命令和结果如下所示：

```
[root@controller ~]# vi /etc/glance/glance-api.conf
[glance_store]
stores = file,http,swift
default_store = swift
filesystem_store_datadir = /var/lib/glance/images/
swift_store_auth_address=http://controller:5000/v3.0
swift_store_endpoint_tyep=internalURL
swift_store_multi_tenant=True
swift_store_admin_tenants=service
swift_store_user=glance
swift_store_key=000000
swift_store_container=chinaskill_glance
swift_store_create_container_on_put=True
swift_store_large_object_size=5120
swift_store_large_object_chunk_size=200
swift_store_region=RegionOne
```

（2）配置完成之后，使用命令重新启动 Glance 组件包含的所有服务，命令如下所示：

```
[root@controller ~]# systemctl restart openstack-glance-*
```

（3）使用 Glance 命令将本地磁盘中内容的进行上传测试，命令如下，结果如图 5-6 所示。

```
[root@controller ~]# glance image-create --name foo --disk-format=qcow2 -
container-format=bare --visibility=public --file /opt/iaas/images/
CentOS_7.5_x86_64_XD.qcow2
```

```
+------------------+--------------------------------------+
| Property         | Value                                |
+------------------+--------------------------------------+
| checksum         | 3d3e9c954351a4b6953fd156f0c29f5c     |
| container_format | bare                                 |
| created_at       | 2021-04-14T11:13:41Z                 |
| disk_format      | qcow2                                |
| id               | 6b097105-63dc-4070-8903-e54df4542e6b |
| min_disk         | 0                                    |
| min_ram          | 0                                    |
| name             | foo                                  |
| owner            | 263606946e8d43d28d094cbea748b9f2     |
| protected        | False                                |
| size             | 510459904                            |
| status           | active                               |
| tags             | []                                   |
| updated_at       | 2021-04-14T11:13:55Z                 |
| virtual_size     | None                                 |
| visibility       | public                               |
+------------------+--------------------------------------+
```

图 5-6　运行结果

（4）使用命令再次查看容器列表，发现存储了一个刚刚创建的镜像，命令和结果如下所示：

```
[root@controller ~]# swift list
chinaskill_glance_6b097105-63dc-4070-8903-e54df4542e6b
```

本任务到此结束。

5.5 开放研究任务：架构 Ceph 分布式存储集群

研发部在使用 GlusterFS 文件系统后，还想再尝试下其他的文件系统，进行对比测试，找出一种性能较好的方式，集成到公司的私有云中。公司决定使用 Ceph 分布式存储集群系统进行文件系统的测试。

任务执行清单

在本任务中，您将通过实践部署 Ceph 分布式存储集群系统，实现基于 Ceph 文件系统的集群存储。

（1）基础环境配置。
（2）Ceph 系统安装。
（3）OSD 和 MDS 部署。
（4）创建 Ceph 文件系统并挂载。

目标

- 掌握 Ceph 分布式存储集群的部署方法。
- 掌握 Ceph 分布式存储集群的基础运维。
- 掌握 Ceph 分布式存储集群的备灾处理方法。

重要信息

- 本任务采用 CentOS 7 操作系统。
- 虚拟机密码为 000000。
- 关闭防火墙和 Selinux。

解决方案

1. 基础环境配置

（1）创建四台虚拟机，节点为 192.168.200.10（20，30，40）。前三台服务器各添加 1 块 20 GB 的硬盘为 sdb，更改主机名，此处以管理节点 ceph-node1 节点为例，

```
[root@ceph-nod1 ~]# hostnamectl set-hostname ceph-node1
[root@ceph-nod1 ~]# logout
[root@ceph-node1 ~]#
```

（2）ceph-node1 做管理 osd，mon 节点，ceph-node2 和 ceph-node3 做 osd，mon，client 为客户端。格式化硬盘，创建相应目录并进行挂载。

```
[root@ceph-node1 ~]# mkfs.xfs /dev/sdb
[root@ceph-node1 ~]# mkdir /var/local/osd{0,1,2}
[root@ceph-node1 ~]# mount /dev/sdb /var/local/osd0
[root@ceph-node1 ceph]# chmod 777  -R /var/local/osd0
```

```
[root@ceph-node2 ~]# mkfs.xfs /dev/sdb
[root@ceph-node2 ~]# mkdir /var/local/osd{0,1,2}
[root@ceph-node2 ~]# mount /dev/sdb /var/local/osd1
[root@ceph-node2 ~]# chmod 777  -R /var/local/osd1

[root@ceph-node3 ~]# mkfs.xfs /dev/sdb
[root@ceph-node3 ~]# mkdir /var/local/osd{0,1,2}
[root@ceph-node3 ~]# mount /dev/sdb  /var/local/osd2/
[root@ceph-node3 ~]# chmod 777  -R /var/local/osd2
```

（3）为了方便起见，修改四台虚拟机的 /etc/hosts 文件，修改主机名地址映射关系。此处以 ceph-node1 为例。

```
[root@ceph-node1 ~]# vi /etc/hosts
127.0.0.1   localhost localhost.localdomain localhost4 localhost4.localdomain4
::1         localhost localhost.localdomain localhost6 localhost6.localdomain6
192.168.200.10 ceph-node1
192.168.200.20 ceph-node2
192.168.200.30 ceph-node3
192.168.200.40 client
```

（4）在 ceph-node1 节点生成 Root SSH 密钥，并将它复制到其他节点上。这些虚拟机的 root（用户）的密码都是 000000。

```
[root@ceph-node1 ~]# ssh-keygen
[root@ceph-node1 ~]# ssh-copy-id  ceph-node1
[root@ceph-node1 ~]# ssh-copy-id  ceph-node2
[root@ceph-node1 ~]# ssh-copy-id  ceph-node3
[root@ceph-node1 ~]# ssh-copy-id  client
```

（5）安装 NTP 服务，使用互联网上提供的 NTP 服务，把 3 台主机时间保持一致（主机需要可以访问互联网）。

```
[root@ceph-node1 ~]# yum -y install ntp
[root@ceph-node1 ~]# ntpdate ntp1.aliyun.com
[root@ceph-node2 ~]# ntpdate ntp1.aliyun.com
[root@ceph-node3 ~]# ntpdate ntp1.aliyun.com
```

（6）增加 yum 配置文件（各节点都要配置 ceph 源），此处以 ceph-node1 节点为例。

```
[root@ceph-node1 ~]# mkdir /etc/yum.repos.d/yum/
[root@ceph-node1 ~]# mv /etc/yum.repos.d/*.repo /etc/yum.repos.d/yum/
[root@ceph-node1 ~]# wget -O /etc/yum.repos.d/CentOS-Base.repo http://mirrors.aliyun.com/repo/Centos-7.repo
[root@ceph-node1 ~]# wget -O /etc/yum.repos.d/epel.repo http://mirrors.aliyun.com/repo/epel-7.repo
[root@ceph-node1 ~]# vi /etc/yum.repos.d/ceph.repo
[ceph]
name=ceph
baseurl=http://mirrors.aliyun.com/ceph/rpm-jewel/el7/x86_64/
gpgcheck=0
priority=1
[ceph-noarch]
```

```
name=cephnoarch
baseurl=http://mirrors.aliyun.com/ceph/rpm-jewel/el7/noarch/
gpgcheck=0
priority=1
[ceph-source]
name=Ceph source packages
baseurl=http://mirrors.aliyun.com/ceph/rpm-jewel/el7/SRPMS/
gpgcheck=0
priority=1
[root@ceph-node1 ~]# yum clean all
[root@ceph-node1 ~]# yum makecache
```

2. Ceph 系统安装

（1）要部署这个集群，需要使用 ceph-deploy 工具在 3 台虚拟机上安装和配置 Ceph。ceph-deploy 是 Ceph 软件定义存储系统的一部分，用来方便地配置和管理 Ceph 存储集群。

```
[root@ceph-node1 ~]# yum -y install ceph-deploy
```

（2）进入配置文件夹中，使用 ceph-deploy 创建一个 ceph 集群，更改配置文件使两个 osd 也能达到 active+clean 状态，命令和结果如下所示：

```
[root@ceph-node1 ~]#  mkdir /etc/ceph && cd /etc/ceph
[root@ceph-node1 ceph]# ceph-deploy new ceph-node1
[root@ceph-node1 ceph]# vi ceph.conf
[global]
fsid = 349b8e1d-33a6-4602-8463-0acb13e9bfab
mon_initial_members = ceph-node1
mon_host = 192.168.200.10
auth_cluster_required = cephx
auth_service_required = cephx
auth_client_required = cephx
osd_pool_default_size = 2
```

ceph-deploy 的 new 子命令能够部署一个默认名称为 Ceph 的新集群，并且它能生成集群配置文件和密钥文件。

（3）在 ceph-node1 节点上，使用 ceph-deploy 工具在所有节点上安装 ceph 二进制软件包。

```
[root@ceph-node1 ceph]#ceph-deploy install ceph-node1 ceph-node2 ceph-node3 client
[root@ceph-node1 ceph]# ceph -v
ceph version 10.2.11 (e4b061b47f07f583c92a050d9e84b1813a35671e)
```

（4）在 ceph-node1 节点上创建 Ceph monitor，并查看 monnitor 状态，命令和结果如下所示。

```
[root@ceph-node1 ceph]# ceph-deploy mon create ceph-node1
[root@ceph-node1 ceph]# ceph-deploy gatherkeys ceph-node1
[root@ceph-node1 ceph]# ceph mon stat
  e1: 1 mons at {ceph-node1=192.168.200.10:6789/0}, election epoch 3, quorum 0 ceph-node1
```

Monitor 创建成功后，这个时候 Ceph 集群并不处于健康状态。下面我们要创建 OSD 服务。

3. OSD 和 MDS 部署

（1）我们实验准备阶段已经创建了目录，并进行硬盘的挂载，此处在 ceph-node1 节点创建 OSD。

```
[root@ceph-node1 ceph]# ceph-deploy osd prepare ceph-node1:/var/local/osd0
```

```
ceph-node2:/var/local/osd1 ceph-node3:/var/local/osd2
```

（2）在 ceph-node1 节点使用 ceph-deploy 工具激活 OSD 节点。

```
[root@ceph-node1 ceph]# ceph-deploy osd activate ceph-node1:/var/local/osd0
ceph-node2:/var/local/osd1 ceph-node3:/var/local/osd2
```

（3）在 ceph-node1 节点使用 ceph-deploy 工具列出 osd 服务三个节点状态。

```
[root@ceph-node1 ceph]# ceph-deploy osd list ceph-node1 ceph-node2 ceph-node3
…省略
[ceph-node1][INFO  ] ceph-0
[ceph-node1][INFO  ] ----------------------------------------
[ceph-node1][INFO  ] Path            /var/lib/ceph/osd/ceph-0
[ceph-node1][INFO  ] ID              0
[ceph-node1][INFO  ] Name            osd.0
[ceph-node1][INFO  ] Status          up
[ceph-node1][INFO  ] Reweight        1.0
[ceph-node1][INFO  ] Active          ok
[ceph-node1][INFO  ] Magic           ceph osd volume v026
[ceph-node1][INFO  ] Whoami          0
[ceph-node1][INFO  ] Journal path    /var/local/osd0/journal
…省略
```

使用 ceph osd tree 查看目录树，结果如图 5-7 所示。

```
[root@ceph-node1 ceph]# ceph osd tree
ID WEIGHT  TYPE NAME           UP/DOWN REWEIGHT PRIMARY-AFFINITY
-1 0.05846 root default
-2 0.01949     host ceph-node1
 0 0.01949         osd.0            up  1.00000          1.00000
-3 0.01949     host ceph-node2
 1 0.01949         osd.1            up  1.00000          1.00000
-4 0.01949     host ceph-node3
 2 0.01949         osd.2            up  1.00000          1.00000
```

图 5-7　运行结果

（4）使用 ceph-deploy 把配置文件和 admin 密钥复制到所有节点，这样每次执行 Ceph 命令行时就无须指定 monitor 地址和 ceph.client.admin.keyring 了，开放权限给其他节点，进行灾备处理。

```
[root@ceph-node1 ceph]# ceph-deploy admin ceph-node1 ceph-node2 ceph-node3
[root@ceph-node1 ceph]# chmod +r /etc/ceph/ceph.client.admin.keyring
```

（5）查看 osd 状态，使用 ceph health 或 ceph -s 命令，结果如下所示：

```
[root@ceph-node1 ceph]# ceph health
HEALTH_OK
[root@ceph-node3 ~]# ceph -s
    cluster e26643b4-f3eb-4dfb-a81a-60e887c1a34e
     health HEALTH_OK
     monmap e1: 1 mons at {ceph-node1=192.168.200.10:6789/0}
            election epoch 3, quorum 0 ceph-node1
     osdmap e14: 3 osds: 3 up, 3 in
            flags sortbitwise,require_jewel_osds
      pgmap v24: 64 pgs, 1 pools, 0 bytes data, 0 objects
            15681 MB used, 45728 MB / 61410 MB avail
                  64 active+clean
```

（6）部署 mds 服务，在 ceph-node1 节点使用 ceph-deploy 管理工具创建两个服务。

```
[root@ceph-node1 ceph]# ceph-deploy mds create ceph-node2 ceph-node3
```

然后使用命令查看服务状态和集群状态

```
[root@ceph-node1 ceph]# ceph mds stat
e3:, 2 up:standby
[root@ceph-node1 ceph]# ceph -s
    cluster e26643b4-f3eb-4dfb-a81a-60e887c1a34e
     health HEALTH_OK
     monmap e1: 1 mons at {ceph-node1=192.168.200.10:6789/0}
            election epoch 3, quorum 0 ceph-node1
     osdmap e14: 3 osds: 3 up, 3 in
            flags sortbitwise,require_jewel_osds
      pgmap v24: 64 pgs, 1 pools, 0 bytes data, 0 objects
            15681 MB used, 45728 MB / 61410 MB avail
                  64 active+clean
```

4. 创建 ceph 文件系统并挂载

（1）ceph-node1 节点需要创建 ceph 文件系统的存储池，在创建之前先查看一下文件系统。

```
[root@ceph-node1 ceph]# ceph fs ls
No filesystems enabled
```

（2）使用相关命令创建两个存储池：cephfs_data 和 cephfs_metadata。命令和结果如下所示：

```
[root@ceph-node1 ceph]# ceph osd pool create cephfs_data 128
pool 'cephfs_data' created
[root@ceph-node1 ceph]#  ceph osd pool create cephfs_metadata 128
pool 'cephfs_metadata' created
```

因为不能自动计算，确定 pg_num 取值是强制性的。下面是几个常用的值：

① 少于 5 个 OSD 时可把 pg_num 设置为 128。

② OSD 数量在 5 ~ 10 个时，可把 pg_num 设置为 512。

③ OSD 数量在 10 ~ 50 个时，可把 pg_num 设置为 4 096。

④ OSD 数量大于 50 时，需要理解权衡方法，自己计算 pg_num 取值。自己计算 pg_num 取值时可借助 pgcalc 工具。

随着 OSD 数量的增加，正确的 pg_num 取值变得更加重要，因为它显著影响着集群的行为以及出错时的数据持久性（即灾难性事件导致数据丢失的概率）。

（3）创建好存储池后，就可以用 fs new 命令创建文件系统了。命令和结果如下所示。

```
[root@ceph-node1 ceph]# ceph fs new 128 cephfs_metadata cephfs_data
new fs with metadata pool 2 and data pool 1
```

（4）查看创建后的 cephfs 和 mds 节点状态。

```
[root@ceph-node1 ceph]# ceph fs ls
name: 128, metadata pool: cephfs_metadata, data pools: [cephfs_data ]
[root@ceph-node1 ceph]# ceph mds stat
e6: 1/1/1 up {0=ceph-node3=up:active}, 1 up:standby
```

mds 状态结果中 active 是活跃的，另 1 个是处于热备份的状态。

（5）挂载 Ceph 文件系统，首先将 ceph-node1 节点中的存储秘钥复制到 Client 客户端的 Ceph 配置文件下（admin.secret 需要自己创建）。

```
[root@ceph-node1 ceph]# cat /etc/ceph/ceph.client.admin.keyring
[client.admin]
        key = AQBGenVg7UORJhAAHEa4b2uvyUKCczNNlyiKLA==
[root@client ~]# vi /etc/ceph/admin.secret
AQBGenVg7UORJhAAHEa4b2uvyUKCczNNlyiKLA==
```

（6）登录 Client 客户端节点，创建一个挂载点，进入相应的配置目录进行 Ceph 文件系统挂载。

```
[root@client ~]# mkdir /opt/ceph
[root@client ~]# cd /etc/ceph/
[root@client ceph]# mount -t ceph 192.168.200.10:6789:/  /opt/ceph/ -o name=admin,secretfile=/etc/ceph/admin.secret
```

（7）在 Client 客户端使用 df -h 命令查看挂载详情，命令和结果如下所示：

```
[root@client ceph]# df -h
Filesystem                Size  Used Avail Use% Mounted on
/dev/mapper/centos-root    36 G  1.3 G  35 G   4% /
devtmpfs                  479 M     0 479 M   0% /dev
tmpfs                     489 M     0 489 M   0% /dev/shm
tmpfs                     489 M  6.7 M 483 M   2% /run
tmpfs                     489 M     0 489 M   0% /sys/fs/cgroup
/dev/sda1                 497 M  114 M 384 M  23% /boot
tmpfs                      98 M     0  98 M   0% /run/user/0
192.168.200.10:6789:/      60 G   16 G  45 G  26% /opt/ceph
```

Ceph 在开源社区还是比较热门的，但是更多的是应用于云计算的后端存储。所以大多数在生产环境中使用 Ceph 的公司都会有专门的团队对 Ceph 进行二次开发，Ceph 的运维难度也比较大。但是经过合理的优化之后，Ceph 的性能和稳定性都是值得期待的。

本任务到此结束。

小　　结

在本章中，您已经学会：
- 共享存储和分布式存储的概念。
- NFS 共享存储部署。
- GlusterFS 分布式存储部署。
- Ceph 分布式服务安装。
- Swift 分布式存储部署。
- Ceph 分布式存储集群部署。

第 6 章

集群与高可用技术

本章概要

学习 Linux 系统中集群架构及高可用技术，提高服务运维的稳定性，部署 Linux 部署集群架构之负载均衡。

学习目标

- 学习 Linux 集群和高可用技术。
- 掌握 Keepalived 服务安装与部署。
- 掌握 LVS 服务安装与部署。
- 掌握 HAProxy 服务安装与部署。
- 掌握 Linux 集群架构之负载均衡。

思维导图

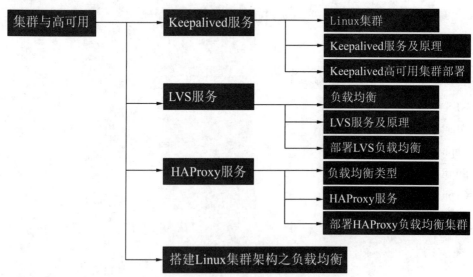

任务目标

学习 Linux 系统中集群架构及高可用技术，提高服务运维的稳定性，部署 Linux 部署集群架构之负载均衡。

6.1 Keepalived 服务

学习目标

学完本节后,您应能够:
- 了解 Linux 集群。
- 了解 Keepalived 服务及原理。
- 掌握 Keepalived 服务部署。

6.1.1 Linux 集群

Linux 集群从功能上可以分为两大类:高可用集群和负载均衡集群。

(1)高可用集群通常为两台服务器,一台工作,一台冗余。当提供服务的服务器宕机时,冗余服务器将接替宕机的服务器继续提供服务。

实现高可用集群的开源软件有 Heartbeat 和 Keepalived。

(2)负载均衡集群,需要一台服务器作为分发器,负责把用户的请求分发给后端服务器处理。在负载均衡集群中,除了分发器外,还有给用户提供服务的服务器,所以,负载均衡集群的服务器最少也需要两台。

实现负载均衡集群的开源软件有很多,如 LVS、Keepalived、HAProxy、Nginx,商业版有 F5 和 Netscaler。商业版的使用费用较高,但并发量高、稳定性好。

6.1.2 Keepalived 服务及原理

1. Keepalived 简介

Keepalived 是集群管理中保证集群高可用的一个服务软件,其功能类似于 heartbeat,用来防止单点故障。Keepalived 是以 VRRP 协议为实现基础的,VRRP 全称 Virtual Router Redundancy Protocol,即虚拟路由冗余协议。

虚拟路由冗余协议,可以认为是实现路由器高可用的协议,即将 N 台提供相同功能的路由器组成一个路由器组,这个组里面有一个 master 和多个 backup,master 上面有一个对外提供服务的 VIP(该路由器所在局域网内其他机器的默认路由为该 VIP),master 会发组播,当 backup 收不到 VRRP 包时就认为 master 宕掉了,这时就需要根据 VRRP 的优先级来选举一个 backup 当 master。这样的话就可以保证路由器的高可用了。

2. VRRP 协议与工作原理

(1)VRRP 协议。

在现实的网络环境中。主机之间的通信都是通过配置静态路由或者(默认网关)来完成的,而主机之间的路由器一旦发生故障,通信就会失效,因此这种通信模式当中,路由器就成了一个单点瓶颈,为了解决这个问题,就引入了 VRRP 协议。

VRRP 协议是一种容错的主备模式的协议,保证当主机的下一跳路由出现故障时,由另一台路由器来代替出现故障的路由器进行工作,通过 VRRP 可以在网络发生故障时透明地进行设备切换而不影响主机之间的数据通信。

① 虚拟路由器:虚拟路由器是 VRRP 备份组中所有路由器的集合,它是一个逻辑概念,并

不是真正存在的。从备份组外面看备份组中的路由器，感觉组中的所有路由器都一样，可以理解为在一个组中：主路由器+所有备份路由器=虚拟路由器。虚拟路由器有一个虚拟的 IP 地址和 MAC 地址。主机将虚拟路由器当作默认网关。虚拟 MAC 地址的格式为 00-00-5E-00-01-{VRID}。通常情况下，虚拟路由器回应 ARP 请求使用的是虚拟 MAC 地址，只有虚拟路由器做特殊配置的时候，才回应接口的真实 MAC 地址。

② 主路由器（master）：虚拟路由器通过虚拟 IP 对外提供服务，而在虚拟路由器内部同一时间只有一台物理路由器对外提供服务，这台提供服务的物理路由器被称为主路由器。一般情况下 Master 是由选举算法产生，它拥有对外服务的虚拟 IP，提供各种网络功能，如 ARP 请求、ICMP 数据转发等。

③ 备份路由器（backup）：虚拟路由器中的其他物理路由器不拥有对外的虚拟 IP，也不对外提供网络功能，仅接受 master 的 VRRP 状态通告信息，这些路由器被称为备份路由器。当主路由器失败时，处于 backup 角色的备份路由器将重新进行选举，产生一个新的主路由器进入 master 角色，继续提供对外服务，整个切换对用户来说是完全透明的。

（2）VRRP 工作过程。

路由器使用 VRRP 功能后，会根据优先级确定自己在备份组中的角色。优先级高的路由器成为 master 路由器，优先级低的成为 backup 路由器。master 拥有对外服务的虚拟 IP，提供各种网络功能，并定期发送 VRRP 报文，通知备份组内的其他设备自己工作正常；backup 路由器只接收 master 发来的报文信息，用来监控 master 的运行状态。当 master 失效时，backup 路由器进行选举，优先级高的 backup 将成为新的 master。

在抢占方式下，当 backup 路由器收到 VRRP 报文后，会将自己的优先级与报文中的优先级进行比较。如果大于通告报文中的优先级，则成为 master 路由器；否则将保持 backup 状态。

在非抢占方式下，只要 master 路由器没有出现故障，备份组中的路由器始终保持 master 或 backup 状态，backup 路由器即使随后被配置了更高的优先级也不会成为 master 路由器。

如果 backup 路由器的定时器超时后仍未收到 master 路由器发送来的 VRRP 报文，则认为 Master 路由器已经无法正常工作，此时 backup 路由器会认为自己是 master 路由器，并对外发送 VRRP 报文。备份组内的路由器根据优先级选举出 master 路由器，承担报文的转发功能。

3. Keepalived 工作原理

Keepalived 工作在 TCP/IP 参考模型的三层、四层、五层（网络层、传输层、应用层）：

（1）网络层（三层）：Keepalived 通过 ICMP 协议向服务器集群中的每一个节点发送一个 ICMP 数据包（有点类似与 Ping 的功能），如果某个节点没有返回响应数据包，那么认为该节点发生了故障，Keepalived 将报告这个节点失效，并从服务器集群中剔除故障节点。

（2）传输层（四层）：Keepalived 在传输层里利用了 TCP 协议的端口连接和扫描技术来判断集群节点的端口是否正常，比如对于常见的 Web 服务器 80 端口。或者 SSH 服务器 22 端口，Keepalived 一旦在传输层探测到这些端口号没有数据响应和数据返回，就认为这些端口发生异常，然后强制将这些端口所对应的节点从服务器集群中剔除。

（3）应用层（五层）：Keepalived 的运行方式更加全面化和复杂化，用户可以通过自定义 Keepalived 工作方式。例如，可以通过编写程序或者脚本来运行 Keepalived，而 Keepalived 将根据用户的设定参数检测各种程序或者服务是否允许正常，如果 Keepalived 的检测结果和用户设定的不一致时，Keepalived 将把对应的服务器从服务器集群中剔除。

4. Keepalived 体系结构

Keepalived 最初是为 LVS 设计的，由于 Keepalived 可以实现对集群节点的状态检测，而 IPVS 可以实现负载均衡功能，因此，Keepalived 借助于第三方模块 IPVS 就可以很方便地搭建一套负载均衡系统。在 Keepalived 当中 IPVS 模块是可配置的，如果需要负载均衡功能，可以在编译 Keepalived 时开打负载均衡功能，也可以通过编译参数关闭。

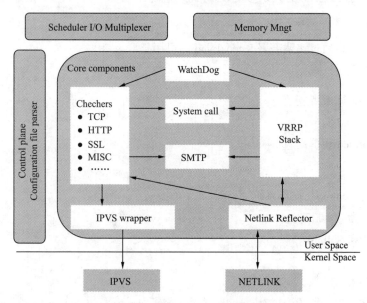

图 6-1　Keepalived 体系结构图

如图 6-1 所示，Scheduler I/O Multiplexer 是一个 I/O 复用分发调度器，它负责安排 Keepalived 所有内部的任务请求；Memory Mngt 是一个内存管理机制，这个框架提供了访问内存的一些通用方法；Control Plane 是 keepalived 的控制版面，可以实现对配置文件的编译和解析；Core componets 主要包含了 5 个部分：

（1）WatchDog：是计算机可靠领域中极为简单又非常有效的检测工具，Keepalived 正是通过它监控 Checkers 和 VRRP 进程的。

（2）Checkers: 是 Keepalived 最基础的功能，也是最主要的功能，可以实现对服务器运行状态检测和故障隔离。

（3）VRRP Stack: VRRP 可以实现 HA 集群中失败切换功能，负责负载均衡器之间的失败切换 FailOver。

（4）IPVS wrapper: 根据自定义配置文件生成相应的 IPVS 规则，IPVS warrper 模块将设置好的 IPVS 规则发送到内核空间，并且提供给 IPVS 模块使用，最终实现 IPVS 模块的负载功能。

（5）Netlink Reflector：用来实现高可用集群 Failover 时虚拟 IP（VIP）的设置和切换，Keepalived 也是模块化设计，不同模块功能也不同，它的三个核心模块分别是 core、check 和 VRRP。

① core 模块：为 Keepalived 的核心组件，负责主进程的启动、维护以及全局配置文件的加载和解析。

② check：负责健康检查，包括常见的各种检查方式。

③ VRRP 模块：用来实现 VRRP 协议。

6.1.3 任务：部署 Keepalived 高可用集群

Keepalived 的作用是检测服务器的状态，如果有一台 Web 服务器宕机，或工作出现故障，Keepalived 将检测到，并将有故障的服务器从系统中剔除，同时使用其他服务器代替该服务器的工作，当服务器工作正常后 Keepalived 自动将服务器加入服务器群中，这些工作全部自动完成，不需要人工干涉，需要人工做的只是修复故障的服务器。

任务执行清单

在本任务中，您将通过实践部署 Keepalived 高可用集群。

目标

- 掌握 Keepalived 服务安装。
- 掌握 Keepalived 服务运维。
- 掌握 Keepalived 集群配置。

重要信息

- 本任务采用 CentOS 7 操作系统。
- 虚拟机密码为 000000。
- 关闭防火墙和 Selinux。

解决方案

1. 基础环境配置

（1）集群准备工作，更改主机名，挂载 centos7 镜像到 /mnt 目录下，将 lnmp_nosql_zaz.tar.gz 包上传至 root 目录下，并解压到 /root/repo 目录下，配置 yum 源码。此处以 master 为例，具体操作如下所示：

```
[root@localhost ~]# hostnamectl set-hostname master
[root@localhost ~]# logout
[root@master ~]# vi /etc/hosts
127.0.0.1      localhost localhost.localdomain localhost4 localhost4.localdomain4
::1            localhost localhost.localdomain localhost6 localhost6.localdomain6
192.168.200.10 master
192.168.200.20 backup
[root@master ~]# mkdir /root/repo/
[root@master ~]# tar -zxvf lnmp_nosql_zaz.tar.gz -C /root/repo/
[root@master ~]# rm -rf /etc/yum.repos.d/*
[root@master ~]# vi /etc/yum.repos.d/local.repo
[centos]
name=centos7
baseurl=file:///mnt
enabled=1
gpgcheck=0
[nginx]
name=nginx
baseurl=file:///root/repo
enabled=1
gpgcheck=0
```

（2）分别在两台虚拟机上安装 Keepalived 和 Nginx 服务，安装服务前需关闭防火墙和 selinux，两台虚拟机都需要执行。此处以 master 为例，具体命令如下所示：

```
[root@master ~]# yum install -y keepalived nginx
[root@master ~]# systemctl start nginx
[root@master ~]# systemctl enable nginx
Created symlink from /etc/systemd/system/multi-user.target.wants/nginx.
service to /usr/lib/systemd/system/nginx.service.
```

2. 设置 Keepalived 主服务器

（1）Keepalived 主配置文件的名称为 /etc/keepalived/keepalived.conf，配置文件主要分为全局配置块、VRRP 配置块，每个配置块都以 {} 包裹，编辑 master 节点上的 keepalived 配置文件，具体修改如下所示：

```
[root@master ~]# vi /etc/keepalived/keepalived.conf
! Configuration File for keepalived
global_defs {
   notification_email {
     acassen@firewall.loc
     failover@firewall.loc
     sysadmin@firewall.loc
   }
   notification_email_from Alexandre.Cassen@firewall.loc
   smtp_server 127.0.0.1                    #修改为本地服务
   smtp_connect_timeout 30
   router_id LVS_DEVEL
}
vrrp_script chk_nginx {                     #增加一个Nginx脚本路径
    script "/usr/local/sbin/check_ng.sh"
    interval 3
}
vrrp_instance VI_1 {
    state MASTER
    interface eno16777736                   #更改网卡名称
    virtual_router_id 51
    priority 100
    advert_int 1
    authentication {
        auth_type PASS
        auth_pass 1111
    }
    virtual_ipaddress {
        192.168.200.30                      #设置虚拟IP
    }
    track_script {
        chk_nginx
    }
}
…省略
```

（2）编写一个用于检测 Nginx 服务是否正常的脚本，内容如下所示：

```
[root@master ~]# vi /usr/local/sbin/check_ng.sh
#!/bin/bash
#author:baicai
d=`date --date today +%Y%m%d_%H:%M%S`      #时间变量，用于记录日志
n=`ps -C nginx --no-heading|wc -l`          #计算nginx进程数量
```

```
# 如果进程数为 0，则启动 Nginx，并且再次检测 Nginx 进程数
# 如果进程数还为 0，则说明 Nginx 无法启动，此时需要关闭 Keepalived
if [ $n -eq 0 ]; then
   systemctl start nginx
   n2=`ps -C nginx --no-heading|wc -l`
   if [ $n2 -eq 0 ]; then
      echo "$d nginx down,keepalived will stop">>/var/log/check_ng.log
      systemctl stop keepalived
   fi
fi
```

（3）更改脚本的权限，使得主机具有相应权限，命令如下所示：

```
[root@master ~]# chmod 755 /usr/local/sbin/check_ng.sh
```

（4）启动 Keepalived 服务，使用 ps aux 查看 Keepalived 进程，检查 Nginx 是否启动，关闭 Nginx，再次查询它时会自动启动，说明脚本已经生效，具体如下所示：

```
[root@master ~]# systemctl start keepalived
[root@master ~]# systemctl stop nginx
[root@master ~]# ps aux | grep nginx
root      24371  0.0  0.1 105436  1984 ?        Ss   10:58   0:00 nginx: master process /usr/sbin/nginx
nginx     24372  0.0  0.2 105904  2964 ?        S    10:58   0:00 nginx: worker process
root      24383  0.0  0.0 112644   952 pts/1    S+   10:58   0:00 grep --color=auto nginx
```

（5）使用命令查看虚拟 IP，可以看见 VIP 在主服务器 master 节点上，结果如图 6-2 所示。

```
[root@master ~]# ip addr
1: lo: <LOOPBACK,UP,LOWER_UP> mtu 65536 qdisc noqueue state UNKNOWN
    link/loopback 00:00:00:00:00:00 brd 00:00:00:00:00:00
    inet 127.0.0.1/8 scope host lo
       valid_lft forever preferred_lft forever
    inet6 ::1/128 scope host
       valid_lft forever preferred_lft forever
2: eno16777736: <BROADCAST,MULTICAST,UP,LOWER_UP> mtu 1500 qdisc pfifo_fast state UP qlen 1000
    link/ether 00:0c:29:8a:01:b2 brd ff:ff:ff:ff:ff:ff
    inet 192.168.200.10/24 brd 192.168.200.255 scope global eno16777736
       valid_lft forever preferred_lft forever
    inet 192.168.200.30/32 scope global eno16777736
       valid_lft forever preferred_lft forever
    inet6 fe80::20c:29ff:fe8a:1b2/64 scope link
       valid_lft forever preferred_lft forever
```

图 6-2 查看网卡信息

3. 设置 Keepalived 从服务器

（1）登录到从节点，编辑 backup 节点上的 Keepalived 配置文件，具体修改如下所示：

```
[root@backup ~]# vi /etc/keepalived/keepalived.conf
! Configuration File for keepalived
global_defs {
   notification_email {
     acassen@firewall.loc
     failover@firewall.loc
     sysadmin@firewall.loc
   }
   notification_email_from Alexandre.Cassen@firewall.loc
   smtp_server 127.0.0.1                      # 修改为本地服务
   smtp_connect_timeout 30
   router_id LVS_DEVEL
```

```
}
vrrp_script chk_nginx {                    #增加一个nginx脚本路径
    script "/usr/local/sbin/check_ng.sh"
    interval 3
}

vrrp_instance VI_1 {
    state BACKUP
    interface eno16777736                  # 更改网卡名称
    virtual_router_id 51
    priority 90                            # 在相同的VRID组中优先级高的为主设备
    advert_int 1
    authentication {
        auth_type PASS
        auth_pass 1111
    }
    virtual_ipaddress {
        192.168.200.30                     #设置虚拟IP
    }
    track_script {
        chk_nginx
    }
}
…省略
```

（2）编写一个用于检测Nginx服务是否正常的脚本，内容如下所示：

```
[root@backup ~]#  vi /usr/local/sbin/check_ng.sh
#!/bin/bash
#author:baicai
d=`date --date today +%Y%m%d_%H:%M%S`       #时间变量,用于记录日志
n=`ps -C nginx --no-heading|wc -l`          #计算nginx进程数量
#如果进程数为0,则启动Nginx,并且再次检测Nginx进程数
#如果进程数还为0,则说明Nginx无法启动,此时需要关闭Keepalived
if [ $n -eq 0 ]; then
   systemctl start nginx
   n2=`ps -C nginx --no-heading|wc -l`
   if [ $n2 -eq 0 ]; then
      echo "$d nginx down,keepalived will stop">>/var/log/check_ng.log
      systemctl stop keepalived
   fi
fi
```

（3）更改脚本的权限，启动Keepalived服务，使用ps aux查看Keepalived进程，命令和结果如下所示：

```
[root@backup ~]# chmod 755 /usr/local/sbin/check_ng.sh
[root@backup ~]# systemctl start keepalived
[root@backup ~]#  ps aux|grep keepalived
root 22656  0.0  0.1 118048  1336 ?    Ss 11:04   0:00 /usr/sbin/keepalived -D
root 22657  0.0  0.3 120296  3056 ?    S  11:04   0:00 /usr/sbin/keepalived -D
root 22658  0.0  0.2 120168  2356 ?    S  11:04   0:00 /usr/sbin/keepalived -D
root 22680  0.0  0.0 112644   956 pts/1 S+ 11:04   0:00 grep --color=auto keepalived
```

4. 区分主从 Nginx 服务

（1）分别修改主服务器和从服务器中 nginx 网站根目录下的 inde.html 文件，具体操作如下所示：

```
[root@master ~]# vi /usr/share/nginx/html/index.html
<html>
    <head>
        <meta charet="UTF-8">
        <title>Keepalived Master</title>
    </head>
    <body>
        <h1>Keepalived Master</h1>
        <h3>host:192.168.100.10</h3>
    </body>
</html>
```

在浏览器的地址栏中输入 IP：192.168.100.10，就会显示刚才添加的信息，如图 6-3 所示。

图 6-3　访问成功

（2）修改从服务器中 nginx 网站根目录下的 inde.html 文件，具体操作如下所示：

```
[root@backup ~]# vi /usr/share/nginx/html/index.html
<html>
    <head>
        <meta charet="UTF-8">
        <title>Keepalived Backup</title>
    </head>
    <body>
        <h1>Keepalived Backup</h1>
        <h3>host:192.168.100.20</h3>
    </body>
</html>
```

在浏览器的地址栏中输入 IP：192.168.100.20，就会显示刚才添加的信息，如图 6-4 所示。

在浏览器的地址栏中输入 VIP 的地址：192.168.100.30，就会访问主服务的内容，因为目前 VIP 在 master 上，如图 6-5 所示。

图 6-4　访问成功　　　　图 6-5　vip 节点访问成功

5. 测试 Keepalived 高可用

（1）关闭主服务器 master 上的 Keepalived 服务，模拟服务器宕机，再分别查看两个节点上的 VIP，命令如下所示：

```
[root@master ~]# systemctl stop keepalived
```

服务正常关闭之后，使用命令查看主节点的 IP 地址，命令和结果如图 6-6 所示。

图 6-6　查看网卡信息（1）

（2）登录到从节点，使用 ip addr 命令查看从节点的 IP 地址状态，发现多了一个虚拟 IP 为 192.168.200.30 的节点，由此可以看出达到了高可用集群状态，结果如图 6-7 所示。

图 6-7　查看网卡信息（2）

这时，VIP 绑定到了从服务器 backup 节点上，使用浏览器再访问 VIP：192.168.100.30，就会访问从服务器上的内容，结果如图 6-8 所示。

图 6-8　访问成功

本任务到此结束。

6.2　LVS 服务

学习目标

学完本节后，您应能够：
- 了解负载均衡。
- 了解 LVS 服务和原理。
- 掌握 LVS 服务的部署。

6.2.1　负载均衡

负载均衡集群实现目前主流开源软件有 LVS、Keepalived、HAProxy、Nginx 等。其中 LVS 属于

4 层（指的是网络 OSI 7 层模型），Nginx 属于 7 层，HAProxy 既可以认为是 4 层，也可以当作 7 层使用。

Keepalived 的负载均衡功能其实就是 LVS，而 HAProxy 比较特殊，支持 4 层，也支持 7 层。LVS 这种 4 层的负载均衡是可以分发除 80 外的其他端口通信的，如 MySQL 的负载均衡，而 Nginx 仅仅支持 http、https、mail。HAProxy 也支持 MySQL 这种 tcp 的负载均衡。相比较来说，LVS 更稳定，能承受更多的请求，Nginx 更灵活，能实现更多的个性化需求。

6.2.2 LVS 服务及原理

1. LVS 概述

LVS（Linux Virtual Server）即 Linux 虚拟服务器，是由章文嵩博士主导的开源负载均衡项目，目前 LVS 已经被集成到 Linux 内核模块中。该项目在 Linux 内核中实现了基于 IP 的数据请求负载均衡调度方案，终端互联网用户从外部访问公司的外部负载均衡服务器，终端用户的 Web 请求会发送给 LVS 调度器，调度器根据自己预设的算法决定将该请求发送给后端的某台 Web 服务器。例如，轮询算法可以将外部的请求平均分发给后端的所有服务器，终端用户访问 LVS 调度器虽然会被转发到后端真实的服务器，但如果真实服务器连接的是相同的存储，提供的服务也是相同的服务，最终用户不管是访问哪台真实服务器，得到的服务内容都是一样的，整个集群对用户而言都是透明的。最后根据 LVS 工作模式的不同，真实服务器会选择不同的方式将用户需要的数据发送到终端用户，LVS 工作模式分为 NAT 模式、TUN 模式、DR 模式。

为了方便大家探讨 LVS 技术，LVS 社区提供了一个命名的约定，内容见表 6-1。

表 6-1 LVS 命令规则

名　称	缩　写	说　明
虚拟 IP	VIP	VIP 为 Director 用于向客户端计算机提供服务的 IP 地址。比如 www.yunjisuan.com 域名就要解析到 vip 上提供服务
真实 IP 地址	RIP	在集群下面节点上使用的 IP 地址，物理 IP 地址
Director 的 IP 地址	DIP	Director 用于连接内外网络的 IP 地址，物理网卡上的 IP 地址。是负载均衡器上的 IP
客户端主机 IP 地址	CIP	客户端用户计算机请求集群服务器的 IP 地址，该地址用作发送给集群的请求的源 IP 地址

LVS 集群内部的节点称为真实服务器（Real Server），也称集群节点。请求集群服务的计算机称为客户端计算机。

与计算机通常在网上交换数据包的方式相同，客户端计算机，Director 和真实服务器使用 IP 地址彼此进行通信。

不同架构角色命名情况如图 6-9 所示。

LVS 集群负载均衡器接受服务的所有入站客户端计算机请求，并根据调度算法决定哪个集群节点应该处理回复请求。负载均衡器（简称 LB）有时也被称为 LVS Director（简称 Director）。

LVS 虚拟服务器的体系结构如图 6-9 所示，一组服务器通过高速的局域网或者地理分布的广域网相互连接，在他们的前端有一个负载调度器（Load Balancer）。负载调度器能无缝地将网络请求调度到真实服务器上，从而使得服务器集群的结构对客户是透明的，客户访问集群系统提供的网络服务就像访问一台高性能，高可用的服务器一样。客户程序不受服务器集群的影响不需要作任何修改。系统的伸缩性通过在服务集群中透明地加入和删除一个节点来达到，通过检测节点或服务进程故障和正确地重置系统达到高可用性。由于负载调度技术是在 Linux 内核中实现的，故称为 Linux 虚拟服务器（Linux Virtual Server）。

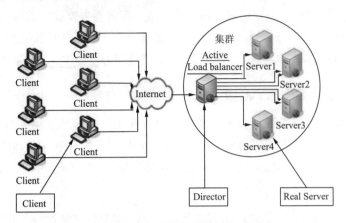

图 6-9　服务角色命名

2. LVS 常见工作模式

（1）基于 NAT 的 LVS 负载均衡。

NAT（Network Address Translation）即网络地址转换，其作用是通过数据报头的修改，使得位于企业内部的私有 IP 地址可以访问外网，以及外部用户可以访问位于公司内部的私有 IP 主机。

（2）基于 TUN 的 LVS 负载均衡。

在 LVS（NAT）模式的集群环境中，由于所有的数据请求及响应的数据包都需要经过 LVS 调度器转发，如果后端服务器的数量大于 10 台，则调度器就会成为整个集群环境的瓶颈。

数据请求包往往远小于响应数据包的大小。因为响应数据包中包含有客户需要的具体数据，所以 LVS（TUN）的思路就是将请求与响应数据分离，让调度器仅处理数据请求，而让真实服务器响应数据包直接返回给客户端。

其中，IP 隧道（IP tunning）是一种数据包封装技术，它可以将原始数据包封装并添加新的包头（内容包括新的源地址及端口、目标地址及端口），从而实现将一个目标为调度器的 VIP 地址的数据包封装，通过隧道转发给后端的真实服务器（Real Server），通过将客户端发往调度器的原始数据包封装，并在其基础上添加新的数据包头（修改目标地址为调度器选择出来的真实服务器的 IP 地址及对应端口），LVS（TUN）模式要求真实服务器可以直接与外部网络连接，真实服务器在收到请求数据包后直接给客户端主机响应数据。

（3）基于 DR 的 LVS 负载均衡。

在 LVS（TUN）模式下，由于需要在 LVS 调度器与真实服务器之间创建隧道连接，这同样会增加服务器的负担。与 LVS（TUN）类似，DR 模式也称直接路由模式，该模式中 LVS 依然仅承担数据的入站请求以及根据算法选出合理的真实服务器，最终由后端真实服务器负责将响应数据包发送返回给客户端。与隧道模式不同的是，直接路由模式（DR 模式）要求调度器与后端服务器必须在同一个局域网内，VIP 地址需要在调度器与后端所有的服务器间共享，因为最终的真实服务器给客户端回应数据包时需要设置源 IP 为 VIP 地址，目标 IP 为客户端 IP，这样客户端访问的是调度器的 VIP 地址，回应的源地址也依然是该 VIP 地址（真实服务器上的 VIP），客户端是感觉不到后端服务器存在的。

由于多台计算机都设置了同一个 VIP 地址，所以在直接路由模式中要求调度器的 VIP 地址是对外可见的，客户端需要将请求数据包发送到调度器主机，而所有的真实服务器的 VIP 地址必须配置在 Non-ARP 的网络设备上,也就是该网络设备并不会向外广播自己的 MAC 及对应的 IP 地址,真实服务器的 VIP 对外界是不可见的,但真实服务器却可以接受目标地址 VIP 的网络请求,并在回应数据包时将源地址设置为该 VIP 地址。调度器根据算法在选出真实服务器后,在不修改数据报文的情况下,将数据帧的 MAC 地址修改为选出的真实服务器的 MAC 地址,通过交换机将该数据帧发给真实服务器。整个过程中,真实服务器的 VIP 不需要对外界可见。

6.2.3 任务:部署 DR 模式的 LVS 负载均衡

网站规模不是很大且服务器在 10 台以内时,可以选择 NAT 模式,因为 NAT 模式节省公网 IP 资源,对于小企业来说,公网 IP 也是需要成本的,所以,用的 IP 少非常占优势。反之,如果几十台、几百台服务器,每一台都设置公网 IP 就非常浪费 IP 资源,尤其是现在公网 IP 越来越少。

任务执行清单

在本任务中,您将通过实践部署基于 DR 模式的 LVS 负载均衡集群服务搭建。

目标

- 掌握 LVS 负载均衡的安装。
- 掌握 LVS 负载均衡的配置。
- 掌握 DR 模式的服务配置。

重要信息

- 本任务采用 CentOS 7 操作系统。
- 虚拟机密码为 000000。
- 提前配置好 yum 源仓库。
- 关闭防火墙和 selinux。

解决方案

1. 基础环境配置

(1)准备 3 台虚拟机,其中一台作为分发器(也称调度器),另外两台作为 Web Server,用作处理用户请求的服务器,首先更改三台虚拟机的主机名。

```
[root@localhost ~]# hostnamectl set-hostname lvs
[root@localhost ~]# logout
[root@lvs ~]#
[root@localhost ~]# hostnamectl set-hostname web2
[root@localhost ~]# logout
[root@web2 ~]#
[root@localhost ~]# hostnamectl set-hostname web3
[root@localhost ~]# logout
[root@web3 ~]#
```

(2)LVS 负载均衡调度器设置,部署网络环境,修改网卡配置文件,命令和结果如下所示:

```
[root@lvs ~]# cd /etc/sysconfig/network-scripts/
[root@lvs network-scripts]# vi ifcfg-eno16777736
TYPE=Ethernet
DEVICE=eno16777736
ONBOOT=yes
IPADDR=192.168.200.10
```

```
NETMASK=255.255.255.0
GATEWAY=192.168.200.1
DNS1=114.114.114.114
[root@lvs network-scripts]# cp ifcfg-eno16777736 ifcfg-eno16777736:1
[root@lvs network-scripts]# vi ifcfg-eno16777736:1
TYPE=Ethernet
BOOTPROTO=static
ONBOOT=yes
IPADDR=192.168.200.11
NETMASK=255.255.255.0
GATEWAY=192.168.200.1
```

重新启动网络信息,并查看信息列表,命令如下,结果如图 6-10 所示。

```
[root@lvs network-scripts]# systemctl restart network
```

```
[root@lvs ~]# ip a
1: lo: <LOOPBACK,UP,LOWER_UP> mtu 65536 qdisc noqueue state UNKNOWN
    link/loopback 00:00:00:00:00:00 brd 00:00:00:00:00:00
    inet 127.0.0.1/8 scope host lo
       valid_lft forever preferred_lft forever
    inet6 ::1/128 scope host
       valid_lft forever preferred_lft forever
2: eno16777736: <BROADCAST,MULTICAST,UP,LOWER_UP> mtu 1500 qdisc pfifo_fast state UP qlen 1000
    link/ether 00:0c:29:8a:01:b2 brd ff:ff:ff:ff:ff:ff
    inet 192.168.200.10/24 brd 192.168.200.255 scope global eno16777736
       valid_lft forever preferred_lft forever
    inet 192.168.200.11/24 brd 192.168.200.255 scope global secondary eno16777736:1
       valid_lft forever preferred_lft forever
    inet6 fe80::20c:29ff:fe8a:1b2/64 scope link
       valid_lft forever preferred_lft forever
```

图 6-10 网卡信息

2. 安装服务

(1)接下来安装 ipvsadm 管理工具,命令如下所示:

```
[root@lvs ~]# yum install -y ipvsadm
```

(2)使用 ipvsadm 管理工具,创建虚拟服务,添加真实服务器组,并为虚拟服务设置适当的调度算法,命令如下所示:

```
[root@lvs ~]# ipvsadm -A -t 192.168.200.11:80 -s rr
[root@lvs ~]# ipvsadm -a -t 192.168.200.11:80 -r 192.168.200.20 -g
#-a 增加服务器,-g 表示 LVS 模式是 DR
[root@lvs ~]# ipvsadm -a -t 192.168.200.11:80 -r 192.168.200.30 -g
#-a 增加服务器,-g 表示 LVS 模式是 DR
```

(3)使用 ipvsadm 管理工具查看规则,显示内核虚拟服务器表,命令和结果如下所示:

```
[root@lvs ~]# ipvsadm -L -n
IP Virtual Server version 1.2.1 (size=4096)
Prot LocalAddress:Port Scheduler Flags
  -> RemoteAddress:Port           Forward Weight ActiveConn InActConn
TCP  192.168.200.11:80 rr
  -> 192.168.200.20:80            Route   1      0          0
  -> 192.168.200.30:80            Route   1      0          0
```

注意:三个 LVS 模式中,只有 NAT 模式需要开启路由转发功能,DR 和 TUN 模式不需要开启。

3. 服务端虚拟机配置

(1)登录到 web2 虚拟机,使用命令更改网络配置信息,命令和结果如下所示:

```
[root@web2 ~]# vi /etc/sysconfig/network-scripts/ifcfg-eno16777736
TYPE=Ethernet
BOOTPROTO=static
ONBOOT=yes
IPADDR=192.168.200.20
```

```
NETMASK=255.255.255.0
GATEWAY=192.168.200.1
DNS1=114.114.114.114
```

（2）增加 web2 回环接口，使用命令生成回环口配置文件，并进行相应的修改，命令和结果如下所示：

```
[root@web2 ~]# vi /etc/sysconfig/network-scripts/ifcfg-lo:0
DEVICE=lo:1
IPADDR=192.168.200.11
NETMASK=255.255.255.255
ONBOOT=yes
NAME=loopback
```

使用命令查看 web2 虚拟机的网络配置信息，如图 6-11 所示。

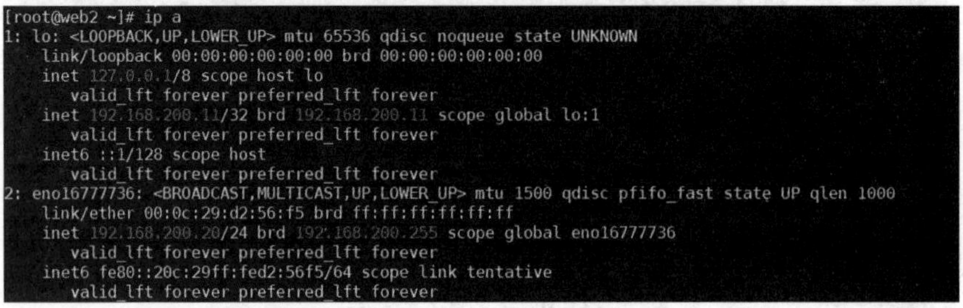

图 6-11　网卡信息

（3）安装 httpd 服务，启动服务之后，设置一个独一无二的网页配置信息，命令和结果如下所示：

```
[root@web2 ~]# yum -y install httpd
[root@web2 ~]#  systemctl restart httpd
[root@web2 ~]# echo "192.168.200.20" > /var/www/html/index.html
```

（4）登录到 web3 虚拟机，使用命令更改网络配置信息，命令和结果如下所示：

```
[root@web3 ~]# vi /etc/sysconfig/network-scripts/ifcfg-eno16777736
TYPE=Ethernet
BOOTPROTO=static
ONBOOT=yes
IPADDR=192.168.200.30
NETMASK=255.255.255.0
GATEWAY=192.168.200.1
DNS1=114.114.114.114
```

（5）增加 web3 回环接口，使用命令生成回环口配置文件，并进行相应的修改，命令和结果如下所示：

```
[root@web3 ~]# vi /etc/sysconfig/network-scripts/ifcfg-lo:0
DEVICE=lo:1
IPADDR=192.168.200.11
NETMASK=255.255.255.255
ONBOOT=yes
NAME=loopback
```

使用命令查看 web3 虚拟机的网络配置信息，如图 6-12 所示。

（6）安装 httpd 服务，启动服务之后，设置一个独一无二的网页配置信息，命令和结果如下所示：

```
[root@web3 ~]# yum -y install httpd
[root@web3 ~]# systemctl restart httpd
[root@web3 ~]# echo "192.168.200.30" > /var/www/html/index.html
```

```
[root@web3 ~]# ip a
1: lo: <LOOPBACK,UP,LOWER_UP> mtu 65536 qdisc noqueue state UNKNOWN
    link/loopback 00:00:00:00:00:00 brd 00:00:00:00:00:00
    inet 127.0.0.1/8 scope host lo
       valid_lft forever preferred_lft forever
    inet 192.168.200.11/32 brd 192.168.200.11 scope global lo:1
       valid_lft forever preferred_lft forever
    inet6 ::1/128 scope host
       valid_lft forever preferred_lft forever
2: eno16777736: <BROADCAST,MULTICAST,UP,LOWER_UP> mtu 1500 qdisc pfifo_fast state UP qlen 1000
    link/ether 00:0c:29:bb:2c:4c brd ff:ff:ff:ff:ff:ff
    inet 192.168.200.30/24 brd 192.168.200.255 scope global eno16777736
       valid_lft forever preferred_lft forever
    inet6 fe80::20c:29ff:febb:2c4c/64 scope link
       valid_lft forever preferred_lft forever
```

图 6-12　网卡信息

（7）关闭 ARP 转发（在两台 RS 上都需要同样的操作），同一个广播域不允许配置多个相同的 VIP，要想实现，就必须让外面的网络无法发现这个 VIP 的存在。此处以 web2 为例。

```
[root@web2 ~]# echo 1 > /proc/sys/net/ipv4/conf/eno16777736/arp_ignore
[root@web2 ~]# echo 2 > /proc/sys/net/ipv4/conf/eno16777736/arp_announce
永久生效：（注意 realserver 的实际 link ok 的网卡是不是 ens33）
[root@web2 ~]# vim /etc/sysctl.conf
net.ipv4.conf.eno16777736.arp_ignore = 1
net.ipv4.conf.eno16777736.arp_announce = 2
[root@web2 ~]# sysctl -p
net.ipv4.conf.eno16777736.arp_ignore = 1
net.ipv4.conf.eno16777736.arp_announce = 2
```

4．服务测试

（1）首先在浏览器中分别输入 web2 和 web3 的 IP 地址，查看网页信息是否正确加载，结果如图 6-13 和图 6-14 所示。

图 6-13　web2 网页显示

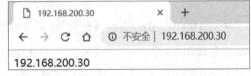
图 6-14　web3 网页显示

（2）客户端使用浏览器访问 http://192.168.200.11，测试 LVS 的 DR 模式，最终可以访问到真实服务器所提供的页面内容，由于 LVS 采用轮训算法，不同的连接请求被分配到不同的后端服务器上，查看网页信息，并不断刷新页面信息，可以看到不同的信息显示，如图 6-15 所示。

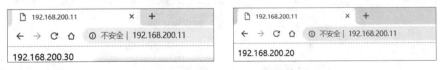
图 6-15　使用 VIP 访问网页

（3）使用 ipvsadm 管理工具查看规则，显示内核虚拟服务器表，可以发现此时两个端点分别访问了 6 次，命令和结果如下所示：

```
[root@lvs ~]# ipvsadm -L -n --stats
IP Virtual Server version 1.2.1 (size=4096)
```

```
Prot LocalAddress:Port           Conns    InPkts   OutPkts   InBytes   OutBytes
  -> RemoteAddress:Port
TCP  192.168.200.11:80             12       667        0     183060        0
  -> 192.168.200.20:80              6       310        0      84810        0
  -> 192.168.200.30:80              6       357        0      98250        0
```

本任务到此结束。

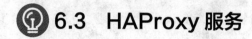 6.3 HAProxy 服务

学习目标

学完本节后，您应能够：
- 了解负载均衡类型。
- 了解什么是 HAProxy。
- 掌握 HAProxy 实现负载均衡集群。

6.3.1 负载均衡类型

1．无负载均衡

没有负载均衡的简单 Web 应用程序环境如图 6-16 所示。

图 6-16 无负载均衡架构

在此示例中，用户直接连接到 Web 服务器 yourdomain.com 上，并且没有负载均衡。如果单个 Web 服务器出现故障，用户将无法再访问 Web 服务器。此外，如果许多用户试图同时访问服务器并且无法处理负载，可能会遇到缓慢的体验，或者可能根本无法连接。

2．4 层负载均衡

将网络流量负载均衡到多个服务器的最简单的方法是使用第 4 层（传输层）负载均衡。以这种方式进行负载均衡将根据 IP 范围和端口转发用户流量（即如果请求进入 http://yourdomain.com/anything，则流量将转发到处理 yourdomain.com 的所有请求的后端），如图 6-17 所示。

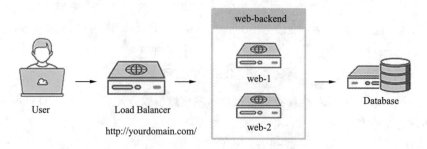

图 6-17 四层负载均衡架构

用户访问负载均衡器，负载均衡器将用户的请求转发给后端服务器 Web 后端组。无论选择哪个后端服务器，都将直接响应用户的请求。通常，Web 后端中的所有服务器应该提供相同的内容，否则用户可能会收到不一致的内容。

3. 7 层负载均衡

更复杂的负载均衡网络流量的方法是使用第 7 层（应用层）负载均衡。使用第 7 层允许负载均衡器根据用户请求的内容将请求转发到不同的后端服务器。这种负载均衡模式允许用户在同一域和端口下运行多个 Web 应用程序服务器，如图 6-18 所示。

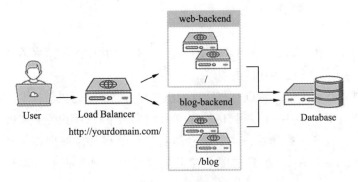

图 6-18　七层负载均衡架构

示例中，如果用户请求 yourdomain.com/blog，则会将其转发到博客后端，后端是一组运行博客应用程序的服务器。其他请求被转发到 web-backend，后端可能正在运行另一个应用程序。

6.3.2　HAProxy 服务

1. HAProxy 概述

（1）HAProxy 是一款提供高可用性、负载均衡以及基于 TCP（第四层）和 HTTP（第七层）应用的代理软件，支持虚拟主机，它是免费、快速并且可靠的一种解决方案。HAProxy 特别适用于那些负载特大的 Web 站点，这些站点通常又需要会话保持或七层处理。HAProxy 运行在当下的硬件上，完全可以支持数以万计的并发连接。并且它的运行模式使得它可以很简单安全地整合进用户当前的架构中，同时可以保护 Web 服务器不被暴露到网络上。

（2）HAProxy 实现了一种事件驱动、单一进程模型，此模型支持非常大的并发连接数。多进程或多线程模型受内存限制、系统调度器限制以及无处不在的锁限制，很少能处理数千并发连接。事件驱动模型因为在有更好的资源和时间管理的用户端（User-Space）实现所有任务，所以没有这些问题。此模型的弊端是，在多核系统上，这些程序通常扩展性较差。这就是为什么必须进行优化以使每个 CPU 时间片（Cycle）做更多的工作。

（3）HAProxy 支持连接拒绝：因为维护一个连接的打开的开销是很低的，有时我们很需要限制攻击蠕虫（attack bots），也就是说限制它们的连接打开，从而限制它们的危害。

（4）HAProxy 支持全透明代理（已具备硬件防火墙的典型特点）：可以用客户端 IP 地址或者任何其他地址来连接后端服务器。这个特性仅在 Linux 2.4/2.6 内核打了 cttproxy 补丁后才可以使用。这个特性也使得为某特殊服务器处理部分流量同时又不修改服务器的地址成为可能。

2. HAProxy 的核心功能

（1）负载均衡：L4 和 L7 两种模式，支持 RR/ 静态 RR/LC/IP Hash/URI Hash/URL_PARAM Hash/HTTP_HEADER Hash 等丰富的负载均衡算法。

（2）健康检查：支持 TCP 和 HTTP 两种健康检查模式。

（3）会话保持：对于未实现会话共享的应用集群，可通过 Insert Cookie/Rewrite Cookie/Prefix Cookie，以及上述的多种 Hash 方式实现会话保持。

（4）SSL：HAProxy 可以解析 HTTPS 协议，并能够将请求解密为 HTTP 后向后端传输。

（5）HTTP 请求重写与重定向。

（6）监控与统计：HAProxy 提供了基于 Web 的统计信息页面，展现健康状态和流量数据。基于此功能，使用者可以开发监控程序来监控 HAProxy 的状态。

3. HAProxy 的关键特性

（1）性能。

① 采用单线程、事件驱动、非阻塞模型，减少上下文切换的消耗，能在 1 ms 内处理数百个请求。并且每个会话只占用数 KB 的内存。

② 大量精细的性能优化，如 O(1) 复杂度的事件检查器、延迟更新技术、Single-buffereing、Zero-copy forwarding 等，这些技术使得 HAProxy 在中等负载下只占用极低的 CPU 资源。

③ HAProxy 大量利用操作系统本身的功能特性，使得其在处理请求时能发挥极高的性能，通常情况下，HAProxy 自身只占用 15% 的处理时间，剩余的 85% 都是在系统内核层完成的。

④ HAProxy 作者在 2009 年使用 1.4 版本进行了一次测试，单个 HAProxy 进程的处理能力突破了 10 万请求 / 秒，并轻松占满了 10 Gbit/s 的网络带宽。

（2）稳定性。

作为建议以单进程模式运行的程序，HAProxy 对稳定性的要求是十分严苛的。按照作者的说法，HAProxy 在十几年间从未出现过一个会导致其崩溃的 BUG，HAProxy 一旦成功启动，除非操作系统或硬件故障，否则不会崩溃。

在上文中提到过，HAProxy 的大部分工作都是在操作系统内核完成的，所以 HAProxy 的稳定性主要依赖于操作系统，作者建议使用 2.6 或 3.x 的 Linux 内核，对 sysctls 参数进行精细的优化，并且确保主机有足够的内存。这样 HAProxy 就能够持续满负载稳定运行数年之久。

（3）GlusterFS 堆栈式结构

采用弱一致性的设计，向副本中写数据时，只要有一个 brick 成功返回，就认为写入成功，不必等待其他 brick 返回，这样的方式比强一致性要快。

6.3.3　任务：部署 HAProxy 负载均衡集群

HAProxy 提供高可用性、负载均衡以及基于 TCP 和 HTTP 应用的代理，支持虚拟主机，它是免费、快速并且可靠的一种解决方案。根据官方数据，其最高支持 10 GB 的并发。

任务执行清单

在本任务中，您将通过实践部署 HAProxy 实现负载均衡集群的部署。

目标

- 掌握 HAProxy 服务安装。
- 掌握七层负载均衡配置。

重要信息

- 本任务采用 CentOS 7 操作系统。
- 虚拟机密码为 000000。
- 提前配置好 yum 源仓库。
- 关闭防火墙和 selinux。

解决方案

1. 基础环境配置

(1) 准备 3 台虚拟机，节点为 192.168.200.10 (20,30)，一台作为 HAProxy 代理服务器，两台作为服务器端，首先更改三台虚拟机的主机名。

```
[root@localhost ~]# hostnamectl set-hostname haproxy
[root@localhost ~]# logout
[root@HAProxy ~]#
[root@localhost ~]# hostnamectl set-hostname web2
[root@localhost ~]# logout
[root@web2 ~]#
[root@localhost ~]# hostnamectl set-hostname web3
[root@localhost ~]# logout
[root@web3 ~]#
```

(2) 添加三台虚拟机 hosts 文件实现集群之间相互解析，此处以 192.168.200.10 节点为例。

```
[root@HAProxy ~]# vi /etc/hosts
127.0.0.1      localhost localhost.localdomain localhost4 localhost4.localdomain4
::1            localhost localhost.localdomain localhost6 localhost6.localdomain6
192.168.200.10 haproxy
192.168.200.20 web2
192.168.200.30 web3
```

2. 安装配置 http 服务

(1) 在 web2 和 web3 虚拟机中安装 http 服务。命令和结果如下所示：

```
[root@web2 ~]# yum install -y httpd
[root@web3 ~]# yum install -y httpd
```

(2) 将测试的网页信息写入 httpd 服务的 index.html 文件中，并启动服务，命令和结果如下所示：

```
[root@web2 ~]# echo "192.168.200.20" > /var/www/html/index.html
[root@web2 ~]# systemctl start httpd
[root@web3 ~]# echo "192.168.200.30" > /var/www/html/index.html
[root@web3 ~]# systemctl start httpd
```

3. 安装配置 HAProxy 服务

(1) 登录到 HAProxy 节点，使用 yum 源安装 HAProxy 服务，命令和结果如下所示：

```
[root@haproxy ~]# yum install -y haproxy
```

(2) 安装好服务之后，配置 HAProxy 代理服务器。具体配置如下所示：

```
[root@haproxy ~]# vi /etc/haproxy/haproxy.cfg
#---------------------------------------------------------------------
# Example configuration for a possible web application.  See the
# full configuration options online.
#
```

```
#       http://haproxy.1wt.eu/download/1.4/doc/configuration.txt
#
#---------------------------------------------------------------------

#---------------------------------------------------------------------
# Global settings
#---------------------------------------------------------------------
global
    # to have these messages end up in /var/log/haproxy.log you will
    # need to:
    #
    # 1) configure syslog to accept network log events.  This is done
    #    by adding the '-r' option to the SYSLOGD_OPTIONS in
    #    /etc/sysconfig/syslog
    #
    # 2) configure local2 events to go to the /var/log/haproxy.log
    #   file. A line like the following can be added to
    #   /etc/sysconfig/syslog
    #
    #    local2.*                       /var/log/haproxy.log
    #
    log                  127.0.0.1 local2

    chroot               /var/lib/haproxy
    pidfile              /var/run/haproxy.pid
    maxconn              4000
    user                 haproxy
    group                haproxy
    daemon

    # turn on stats unix socket
    stats socket /var/lib/haproxy/stats

#---------------------------------------------------------------------
# common defaults that all the 'listen' and 'backend' sections will
# use if not designated in their block
#---------------------------------------------------------------------
defaults
    mode                         http
    log                          global
    option                       httplog
    option                       dontlognull
    option http-server-close
    option forwardfor            except 127.0.0.0/8
    option                       redispatch
    retries                      3
    timeout http-request         10s
    timeout queue                1m
    timeout connect              10s
    timeout client               1m
    timeout server               1m
    timeout http-keep-alive      10s
    timeout check                10s
```

```
        maxconn                     3000
#---------------------------------------------------------------
# main frontend which proxys to the backends
#---------------------------------------------------------------
frontend  main *:80                        #定义终端用户的代理端口
    #acl url_static       path_beg       -i /static /images /javascript /stylesheets
                                           # 禁用其他 URL
    #acl url_static       path_end       -i .jpg .gif .png .css .js
                                           # 禁用其他 URL
    #use_#backend static           if url_static
                                           # 禁用
    default_backend              app
#---------------------------------------------------------------
# static backend for serving up images, stylesheets and such
#---------------------------------------------------------------
#backend static                             # 禁用其他静态后端配置
#balance       roundrobin
    #server       static 127.0.0.1:4331 check
#---------------------------------------------------------------
# round robin balancing between the various backends
#---------------------------------------------------------------
backend app
    balance        roundrobin
    server  web2 192.168.200.20:80 cookie web2 check inter 2000 rise 2 fall 3 weight 1
    # 定义后端服务器,想 cookie 插入 web2 信息,check 代表允许对服务器进行健康检查,检查间隔为 2000ms,连续两次健康检查成功则为服务器有效开启,连续三次健康失败则为服务器宕机,服务器权重为 1
    server  web3 192.168.200.30:80 cookie web2 check inter 2000 rise 2 fall 3 weight 1
    listen admin_stats                     #定义 haproxy 的监控界面
        stats   enable
        bind    *:8080                     # 访问端口为 8080
        mode    http
        option  httplog
        log         global
        maxconn 10
        stats   refresh 30s                # 监控统计页面自动刷新页面时间为 5 s
        stats   uri /admin?stats           # 监控页面 URL 路径
        stats   realm haproxy              # 登录监控统计页面提示符
        stats   auth admin:admin           # 查看 haproxy 监控页面的账户与密码
        stats   hide-version               # 隐藏 haproxy 版本信息
        stats   admin if TRUE
```

(3)启动服务,在浏览器中输入 192.168.200.10,并不断刷新查看是否实现 HAProxy 负载均衡实验配置。结果如图 6-19 所示。

在浏览器中可以看见输入 HAProxy 代理服务器 IP 地址后,刷新将分别得到 web2 和 web3 两台主机返回的页面信息。

(4)在浏览器中访问 http://192.168.200.10:8080/admin?stats,查看代理服务器状态统计信息,如图 6-20 所示,需要输入用户名和密码(admin:admin),登录即可看到图 6-21 所示的状态统计结果。

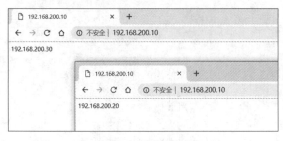

图 6-19　网页测试　　　　　　　　　　　图 6-20　网页登录

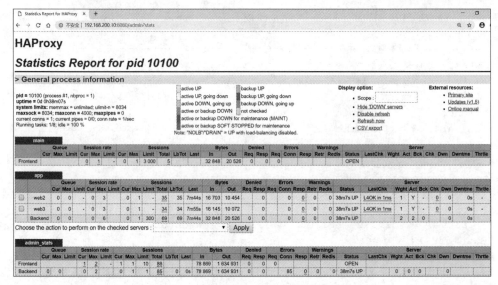

图 6-21　HAProxy 服务器状态统计

本任务到此结束。

6.4　开放研究任务：搭建 Linux 集群架构之负载均衡

我们选择了 Keepalived 来实现高可用，其设计之初就是专为 LVS 负载均衡软件设计的，用来管理并监控 LVS 系统各个服务节点的状态，后来加入了可以实现高可用的 VRRP 功能。因此 Keepalived 除了能管理 LVS 软件外，还可以作为其他服务（例如：Nginx、HAProxy、MySQL 等）的高可用解决方案软件。

任务执行清单

在本任务中，您将通过实践安装 Keepalived、LVS-NAT 服务，搭建 Linux 集群架构之负载均衡。
（1）基础环境配置。
（2）安装并配置服务。
（3）服务测试。

目标

- 掌握 Keepalived 部署和安装。
- 掌握 LVS-NAT 负载均衡部署和安装。
- 掌握 Linux 集群架构部署和负载均衡配置。

第 6 章 集群与高可用技术

📎 重要信息

- 本任务采用 CentOS 7 操作系统。
- 虚拟机密码为 000000。
- 提前配置好 yum 源仓库。
- 关闭防火墙和 Selinux。

✏️ 解决方案

1. 基础环境配置

（1）集群准备工作，创建四台虚拟机，虚拟机集群节点规划如表 6-2 所示。

表 6-2 节点规划

内部 IP	外部 IP	角 色	名 称	备 注
192.168.200.10	192.168.100.10	LVS 负载均衡器（主）	dr1	VIP：192.168.100.100 DIP：192.168.200.100
192.168.200.20	192.168.100.20	LVS 负载均衡器（从）	dr2	VIP：192.168.100.100 DIP：192.168.200.100
192.168.200.30	无	Web01 节点	rs1	
192.168.200.40	无	Web02 节点	rs2	

（2）登录到各个虚拟机，按表格修改虚拟机的主机名，此处以 dr1 为例，命令和结果如下所示：

```
[root@localhost ~]# hostnamectl set-hostname dr1
[root@localhost ~]# logout
[root@dr1 ~]#
```

（3）在两台 Web 服务器上安装 net-tools 包，并设置网关为 192.168.200.100，使用 route -n 命令可以看见网关为 192.168.200.100，此处以 rs1 为例，命令和结果如下所示：

```
[root@rs1 ~]# vi /etc/sysconfig/network-scripts/ifcfg-eno16777736
TYPE=Ethernet
BOOTPROTO=static
DEVICE=eno16777736
ONBOOT=yes
IPADDR=192.168.200.30
NETMASK=255.255.255.0
GATEWAY=192.168.200.100
[root@rs1 ~]# systemctl restart network
[root@rs1 ~]# route -n
Kernel IP routing table
Destination     Gateway         Genmask         Flags Metric Ref    Use Iface
0.0.0.0         192.168.200.100 0.0.0.0         UG    100    0        0 eno16777736
192.168.200.0   0.0.0.0         255.255.255.0   U     100    0        0 eno16777736
```

（4）修改 dr1 和 dr2 服务器的 sysctl 内核参数，通过 net.ipv4.ip_forward = 1 开启网络转发功能，禁用所有网卡的 ICMP 重定向，并使之生效，此处以 dr1 节点为例，命令和结果如下所示：

```
[root@dr1 ~]# vi /etc/sysctl.conf
# System default settings live in /usr/lib/sysctl.d/00-system.conf.
# To override those settings, enter new settings here, or in an /etc/sysctl.d/<name>.conf file
#
```

```
# For more information, see sysctl.conf(5) and sysctl.d(5).
net.ipv4.ip_forward = 1
net.ipv4.conf.all.send_redirects = 0
net.ipv4.conf.default.send_redirects = 0
net.ipv4.conf.eno16777736.send_redirects = 0
net.ipv4.conf.eno33554960.send_redirects = 0
[root@dr1 ~]# sysctl -p
net.ipv4.ip_forward = 1
net.ipv4.conf.all.send_redirects = 0
net.ipv4.conf.default.send_redirects = 0
net.ipv4.conf.eno16777736.send_redirects = 0
net.ipv4.conf.eno33554960.send_redirects = 0
```

2. 安装并配置服务

（1）在 dr1 和 dr2 节点，使用命令安装 Keepalived 服务和 ipvsadm 管理工具，用来设置 LVS 负载均衡器的主和备。

```
[root@dr1 ~]# yum -y install keepalived ipvsadm
[root@dr2 ~]# yum -y install keepalived ipvsadm
```

（2）服务安装好之后，在 dr1 和 dr2 节点根据实际要求配置 Keepalived 服务，具体配置如下所示：

```
[root@dr1 ~]# vi /etc/keepalived/keepalived.conf
! Configuration File for keepalived

global_defs {
   notification_email {
     acassen@firewall.loc
     failover@firewall.loc
     sysadmin@firewall.loc
   }
   notification_email_from Alexandre.Cassen@firewall.loc
   smtp_server 127.0.0.1
   smtp_connect_timeout 30
   router_id LVS_DEVEL
}

vrrp_instance VI_1 {
    state MASTER
    interface eno16777736
    virtual_router_id 51
    priority 100
    advert_int 1
    authentication {
        auth_type PASS
        auth_pass 1111
    }
    virtual_ipaddress {
        192.168.100.100
    }
}
vrrp_instance VI_2 {
    state MASTER
    interface eno16777736
```

```
        virtual_router_id 51
        priority 100
        advert_int 1
        authentication {
            auth_type PASS
            auth_pass 1111
        }
        virtual_ipaddress {
            192.168.200.100
        }
}
virtual_server 192.168.100.100 80 {
    delay_loop 10
    lb_algo wlc
    lb_kind NAT
    net_mask 255.255.255.0
    persistence_timeout 0
    protocol TCP

    real_server 192.168.200.30 80 {
        weight 100
        TCP_CHECK {
            connect_timeout 10
            nb_get_retry 3
            delay_before_retry 3
            connect_port 80
        }
    }

    real_server 192.168.200.40 80 {
        weight 100
        TCP_CHECK {
            connect_timeout 10
            nb_get_retry 3
            delay_before_retry 3
            connect_port 80
        }
    }
}

[root@dr2 ~]# vi /etc/keepalived/keepalived.conf
! Configuration File for keepalived
global_defs {
   notification_email {
     acassen@firewall.loc
     failover@firewall.loc
     sysadmin@firewall.loc
   }
   notification_email_from Alexandre.Cassen@firewall.loc
   smtp_server 127.0.0.1
   smtp_connect_timeout 30
   router_id LVS_DEVEL
}
```

```
vrrp_instance VI_1 {
    state MASTER
    interface eno33554960
    virtual_router_id 52
    priority 98
    advert_int 1
    authentication {
        auth_type PASS
        auth_pass 1111
    }
    virtual_ipaddress {
        192.168.100.100
    }
}
vrrp_instance VI_2 {
    state MASTER
    interface eno16777736
    virtual_router_id 52
    priority 98
    advert_int 1
    authentication {
        auth_type PASS
        auth_pass 1111
    }
    virtual_ipaddress {
        192.168.200.100
    }
}
virtual_server 192.168.100.100 80 {
    delay_loop 10
    lb_algo wlc
    lb_kind NAT
    net_mask 255.255.255.0
    persistence_timeout 0
    protocol TCP

    real_server 192.168.200.30 80 {
        weight 100
        TCP_CHECK {
            connect_timeout 10
            nb_get_retry 3
            delay_before_retry 3
            connect_port 80
        }
    }
    real_server 192.168.200.40 80 {
        weight 100
        TCP_CHECK {
            connect_timeout 10
            nb_get_retry 3
            delay_before_retry 3
            connect_port 80
        }
    }
}
```

（3）在两台 Web 服务器中，安装 Nginx 服务，启动之后设为开机自启动，然后输入一个标识到网页上，命令和结果如下所示：

```
[root@rs1 ~]# yum -y install nginx
[root@rs1 ~]# systemctl start nginx
[root@rs1 ~]# systemctl enable nginx
Created symlink from /etc/systemd/system/multi-user.target.wants/nginx.service to /usr/lib/systemd/system/nginx.service.
[root@rs1 ~]# echo "192.168.200.30 Nginx" > /usr/share/nginx/html/index.html
[root@rs2 ~]# yum -y install nginx
[root@rs2 ~]# systemctl start nginx
[root@rs2 ~]# systemctl enable nginx
Created symlink from /etc/systemd/system/multi-user.target.wants/nginx.service to /usr/lib/systemd/system/nginx.service.
[root@rs2 ~]# echo "192.168.200.40 Nginx" > /usr/share/nginx/html/index.html
```

（4）在两台 LVS 服务器上，启动 Keepalived 服务。在 dr1 节点启动完成之后，使用 ip a 命令可以看到生成的外部 vip：192.168.100.100，内部 DIP 网关为：192.168.200.100，如图 6-22 所示。

```
[root@dr1 ~]# systemctl start keepalived
[root@dr1 ~]# systemctl enable keepalived
Created symlink from /etc/systemd/system/multi-user.target.wants/keepalived.service to /usr/lib/systemd/system/keepalived.service.
```

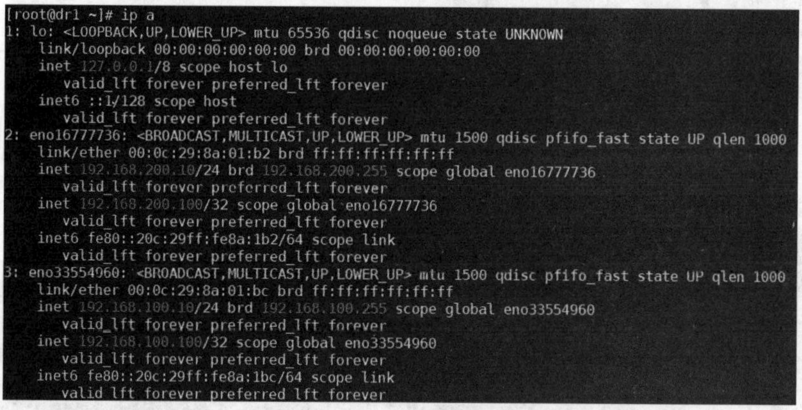

图 6-22　dr1 节点网卡信息

登录到 dr2 节点执行 Keepalived 服务启动命令之后，可以看到图 6-23 所示结果。

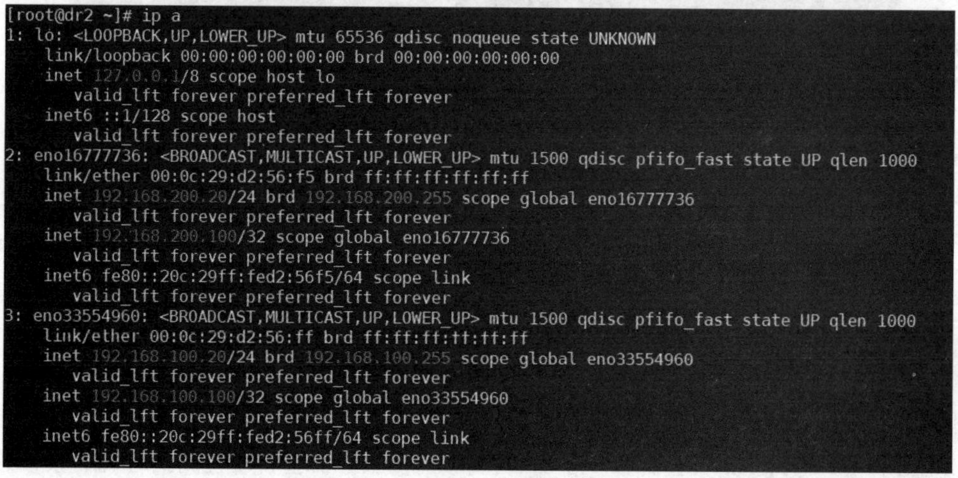

图 6-23　dr2 节点网卡信息

（5）使用 dr1 和浏览器客户端进行访问测试，并在 LVS 主上进行访问连接查询，命令和结果如下所示。在主备服务器上检查 ipvsadm 服务器状态。

```
[root@dr1 ~]# ipvsadm -Ln
IP Virtual Server version 1.2.1 (size=4096)
Prot LocalAddress:Port Scheduler Flags
  -> RemoteAddress:Port           Forward Weight ActiveConn InActConn
TCP  192.168.100.100:80 wlc
  -> 192.168.200.30:80            Masq    100    0          0
  -> 192.168.200.40:80            Masq    100    0          0
```

3. 服务测试

（1）登录 dr1 节点，在该节点中，使用 curl http://192.168.100.100 命令进行网页服务的查询，命令和结果如图 6-24 所示。

（2）打开两个浏览器，并在浏览器中输入 IP：192.168.100.100，可以看到显示两个不同的网页显示结果，如图 6-25 所示。

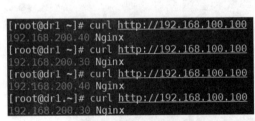

图 6-24　负载均衡测试 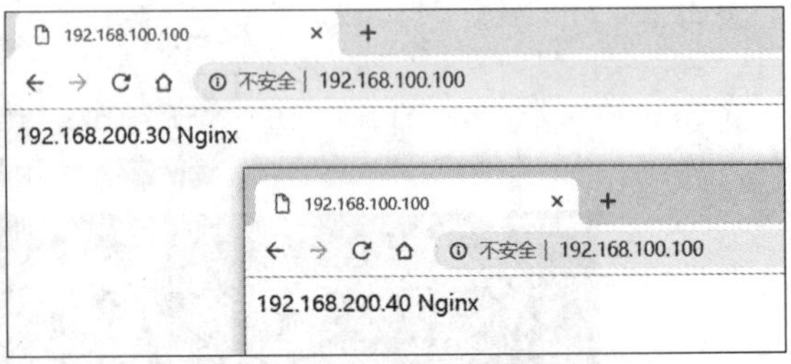　　　　　　　图 6-25　负载均衡网页测试

（3）停掉主服务器的 Keepalived 服务器后，模拟服务器宕机，VIP 的流量就被转移到了 DR2 的机器上，继续提供服务，验证如图 6-26 所示。

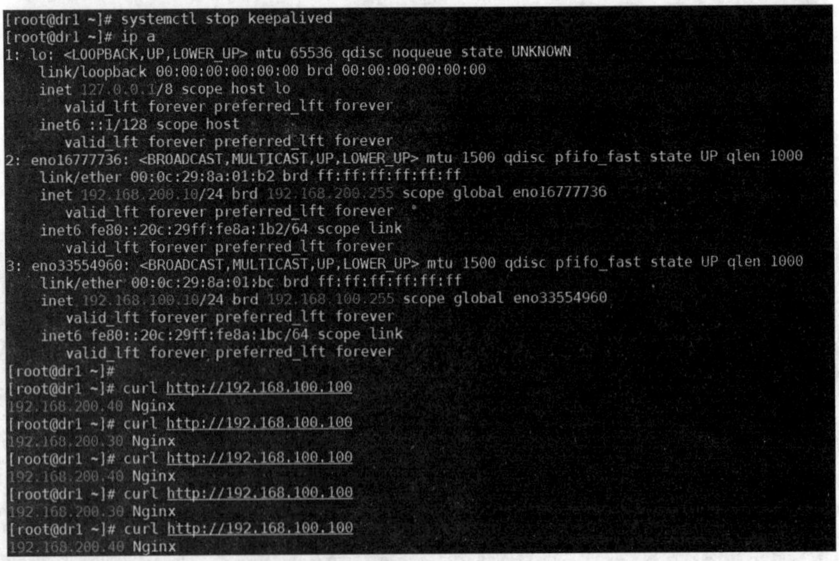

图 6-26　验证结果

根据该结果发现依然可以通过 curl 命令获取网页的返回信息。

（4）停掉 rs1 的 Nginx 服务器后，ipvsadm 负载均衡服务器会自动检测到宕机的服务器，并在负载均衡中删除宕机的服务器，保证前端服务正常访问。

宕机前使用命令查看，如图 6-27 所示。

图 6-27　状态信息查看

关闭 rs1 上的 nginx：

[root@rs1 ~]# systemctl stop nginx

ipvsadm 命令用于检测服务宕机，删除宕机的服务器。此时的高可用集群架构起到了作用，此架构容错性好，不影响客户端访问，结果如图 6-28 所示。

图 6-28　高可用集群验证

本任务到此结束。

小　　结

在本章中，您已经学会：
- Keepalived 服务安装与部署。
- LVS 服务安装与部署。
- HAProxy 服务安装与部署。
- Linux 集群架构之负载均衡。

第 7 章 监控服务与自动化运维工具

本章概要

学习 Linux 操作系统中的 Nagios 和 Zabbix 监控服务的安装与部署,并使用 Ansible 自动化运维服务部署 Nginx 和 LAMP 环境。

学习目标

- 掌握 Nagios 监控服务安装与部署。
- 掌握 Zabbix 监控服务安装与部署。
- 掌握 Ansible 自动化服务安装与部署。
- 掌握 Ansible 批量部署 LAMP 环境。

思维导图

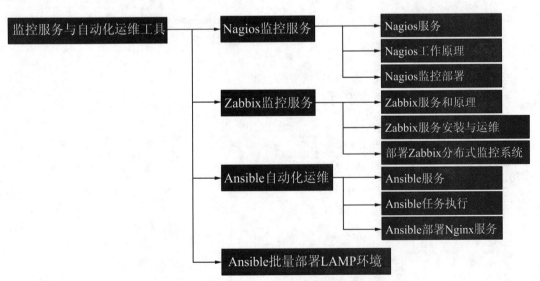

任务目标

学习 Linux 系统中监控服务的安装与部署,使用 Ansible 自动化运维服务进行部署。

7.1 Nagios 监控服务

学习目标

学完本节后，您应能够：
- 了解什么是 Nagios。
- 了解 Nagios 服务及原理。
- 掌握 Nagios 监控案例实施。

7.1.1 Nagios 服务

Nagios 是一款企业级开源免费的监控工具，该工具可以监控应用服务器、交换机、路由器等网络设备，并在服务或设备发生异常时发出报警信息。与 Cacti 不同的是，Nagios 重点专注的是运行在服务器上的服务是否正常，而不会像 Cacti 一样生成图形；Nagios 关注点在于保证服务的正常运行，并且在服务发生问题时提供报警机制。Nagios 强大而灵活的监控预警机制可以帮助企业在出现严重故障前把问题解决，报警信息可以通过邮件或短信发送给管理员，让管理员实时掌控服务器运行状态。部署 Nagios 系统除了需要安装主程序，还需要安装相关插件，Nagios 是整个监控平台的主程序，Nagios-plugins 是必选的插件程序。另外，官方网站还提做了一些可选的附加插件。其中，NRPE 插件用来监控远程 Linux 服务器的主机资源，NSClient 插件用来监控 Windows 主机，NDOUtils 插件需要结合数据库系统将 Nagios 进程的当前数据和历史数据写入数据库。Nagios 可以监控常见的 HTTP、POP3、SMTP、FTP、SSH、PING 等服务，也可以监控主机的 CPU、磁盘等主机资源。如果这些标准的插件无法满足企业中特殊的监控需求，还可以开发自己的监控插件实现特殊的监控功能。

7.1.2 Nagios 工作原理

1. Nagios 工作原理

Nagios 的功能是监控服务和主机，但是它自身并不包括这部分功能，所有的监控、检测功能都是通过各种插件来完成的。

启动 Nagios 后，它会周期性地自动调用插件去检测服务器状态，同时 Nagios 会维持一个队列，所有插件返回来的状态信息都进入队列，Nagios 每次都从队首开始读取信息，并进行处理后，把状态结果通过 Web 显示出来。

Nagios 提供了许多插件，利用这些插件可以方便地监控很多服务状态。安装完成后，在 nagios 主目录下的 /libexec 里放有 nagios 自带的可以使用的所有插件，如，check_disk 是检查磁盘空间的插件，check_load 是检查 CPU 负载的，等等。每一个插件可以通过运行 ./check_xxx –h 来查看其使用方法和功能。

2. Nagios 的 4 种监控状态

Nagios 可以识别 4 种状态返回信息：

（1）0(OK) 表示状态正常 / 绿色。

（2）1(WARNING) 表示出现警告。

（3）2(CRITICAL) 表示出现非常严重的错误 / 红色。

（4）3(UNKNOWN) 表示未知错误 / 深。

Nagios 根据插件返回的值判断监控对象的状态，并通过 Web 显示出来，以便管理员及时发现故障。

3. Nagios 报警插件

对于报警功能，如果监控系统发现问题不能报警就没有意义了，所以报警也是 Nagios 很重要的功能之一。但是，Nagios 自身也没有报警部分的代码，甚至没有插件，而是交给用户或者其他相关开源项目组去完成的。

那么 Nagios 是如何管理远端服务器对象的？Nagios 系统提供了一个插件 NRPE。Nagios 通过周期性的运行这个插件来获得远端服务器的各种状态信息。

Nagios 通过 NRPE 来远端管理服务：

（1）Nagios 执行安装在它里面的 check_nrpe 插件，并告诉 check_nrpe 去检测哪些服务。

（2）通过 SSL，check_nrpe 连接远端机器上的 NRPE daemon。

（3）NRPE 运行本地的各种插件去检测本地的服务和状态 (check_disk,..etc)。

（4）NRPE 把检测的结果传给主机端的 check_nrpe，check_nrpe 再把结果送到 Nagios 状态队列中。

（5）Nagios 依次读取队列中的信息，再把结果显示出来。

7.1.3 Nagios 监控部署

Nagios 本身不包括监控主机和服务，所有功能由插件构成，重点关注于运行在服务器的服务是否正常，不生成图像，提供强大的监控预警机制和服务异常报警。

任务执行清单

在本任务中，您将通过实践部署 Nagios 监控应用案例。

目标

- 掌握 Nagios 服务安装与配置。
- 掌握 Nagios 插件安装与配置。

重要信息

- 本任务采用 CentOS 7 操作系统。
- 虚拟机密码为 000000。
- 关闭防火墙和 Selinux。

1．Nagios 服务安装

（1）基础环境配置，准备 3 台虚拟机，节点为 192.168.200.10（20,30），一台作为 Nagios 服务器，两台作为客户端，首先更改三台虚拟机的主机名。

```
[root@localhost ~]# hostnamectl set-hostname nagios
[root@localhost ~]# logout
[root@nagios ~]#
[root@localhost ~]# hostnamectl set-hostname web2
[root@localhost ~]# logout
[root@web2 ~]#
[root@localhost ~]# hostnamectl set-hostname web3
[root@localhost ~]# logout
[root@web3 ~]#
```

(2)在配置好 yum 源之后,在 Nagios 节点安装所需要的依赖软件包,命令如下所示:

```
[root@nagios ~]# yum install gd gd-devel openssl openssl-devel httpd php gcc
glibc glibc-common make net-snmp wget
```

(3)首先创建账户和组,然后将 nagios-3.5.0.tar.gz 安装包上传至 Nagios 节点的 root 目录下,进行解压。

```
[root@nagios ~]# groupadd nagios
[root@nagios ~]# useradd -g nagios nagios
[root@nagios ~]# tar -zxvf nagios-3.5.0.tar.gz -C /usr/src/
[root@nagios ~]# cd /usr/src/nagios/
```

(4)使用命令进行编译安装,安装相应的插件,最后需要通过多个 make install 命令安装部署不用的监控配置文件与目录,命令和结果如下所示:

```
[root@nagios nagios]# ./configure --with-nagios-user=nagios
--with-nagios-group=nagios
[root@nagios nagios]# make all
[root@nagios nagios]# make install
[root@nagios nagios]# make install-init
[root@nagios nagios]# make install-commandmode
[root@nagios nagios]# make install-config
[root@nagios nagios]# make install-webconf
```

由于 Nagios 最终以 Web 的形式进行管理与监控,安装过程中需要使用 make install-webconf 命令将生成一个 Web 配置文件。

(5)Nagios 拥有众多强大的插件,此处可以根据实际情况进行安装,将 nagios-plugins-2.3.3.tar.gz 安装包上传至 Nagios 节点的 root 目录下,并进行解压,进入目录完成编译安装部署,命令如下所示:

```
[root@nagios ~]# tar -zxvf nagios-plugins-2.3.3.tar.gz -C /usr/src/
[root@nagios ~]# cd /usr/src/nagios-plugins-2.3.3/
[root@nagios nagios-plugins-2.3.3]# ./configure --prefix=/usr/local/nagios
[root@nagios nagios-plugins-2.3.3]# make
[root@nagios nagios-plugins-2.3.3]# make install
```

将 nrpe-2.14.tar.gz 安装包上传至 Nagios 节点的 root 目录下,进行解压之后,完成编译安装,命令如下所示:

```
[root@nagios ~]# tar -zxvf nrpe-2.14.tar.gz -C /usr/src/
[root@nagios ~]# cd /usr/src/nrpe-2.14/
[root@nagios nrpe-2.14]# ./configure
[root@nagios nrpe-2.14]# make all
[root@nagios nrpe-2.14]# make install-plugin
[root@nagios nrpe-2.14]# make install-daemon
[root@nagios nrpe-2.14]# make install-daemon-config
```

(6)设置服务目录的权限,并使用命令更改服务的用户名密码,最后启动服务,命令和结果如下所示:

```
[root@nagios ~]# chown -R nagios:nagios /usr/local/nagios/
[root@nagios ~]# htpasswd -c /usr/local/nagios/etc/htpasswd.users admin
New password:
Re-type new password:
Adding password for user admin
```

```
[root@nagios ~]# systemctl start httpd
[root@nagios ~]# /etc/init.d/nagios start
Reloading systemd:                                    [  OK  ]
Starting nagios (via systemctl):                      [  OK  ]
```

2. Nagios 服务配置

Nagios 配置文件比较多，表 7-1 为主要配置文件及用途，所有的配置文件都位于 /usr/local/nagios/etc 目录及 objects 子目录下。

表 7-1　Nagios 主要配置文件及用途

文件名或目录名	用　　途
cgi.cfg	控制 CGI 访问的配置文件
nagios.cfg	Nagios 主配置文件
resource.cfg	变量定义文件，又称为资源文件，在些文件中定义变量，以便由其他配置文件引用，如 $USER1$
objects	objects 是一个目录，在此目录下有很多配置文件模板，用于定义 Nagios 对象
commands.cfg	命令定义配置文件，其中定义的命令可以被其他配置文件引用
contacts.cfg	定义联系人和联系人组的配置文件
localhost.cfg	定义监控本地主机的配置文件
printer.cfg	定义监控打印机的一个配置文件模板，默认没有启用此文件
switch.cfg	定义监控路由器的一个配置文件模板，默认没有启用此文件
templates.cfg	定义主机和服务的一个模板配置文件，可以在其他配置文件中引用
timeperiods.cfg	定义 Nagios 监控时间段的配置文件
windows.cfg	监控 Windows 主机的一个配置文件模板，默认没有启用此文件

（1）下面根据需求依次修改 Nagios 所有的监控配置文件，以满足本例要求，配置文件中被修改的位置会使用粗体字表示，很多配置文件不需要修改就可以直接使用。

在主配置文件中使用 cfg_file 配置项加载其他配置文件信息，需要监控两台主机设备，为了方便管理，将两台监控主机创建不同的配置文件，在该配置文件中需要加载这些用户自己创建的配置文件。

```
[root@nagios ~]# vi /usr/local/nagios/etc/nagios.cfg
……
# Definitions for monitoring the local (Linux) host
cfg_file=/usr/local/nagios/etc/objects/localhost.cfg
cfg_file=/usr/local/nagios/etc/objects/web2.cfg
cfg_file=/usr/local/nagios/etc/objects/web3.cfg
……
```

（2）修改 CGI 配置文件，将创建的 admin 用户添加至该配置文件中，使得可以正常使用 CGI 程序。

```
# 要求经过验证的用户才可以使用 CGI 程序
[root@nagios ~]# vi /usr/local/nagios/etc/cgi.cfg
# 设置允许通过 CGI 查看 Nagios 进程信息的账户名称
authorized_for_system_information=nagiosadmin,admin
# 设置可以查看所有配置信息的账户名称
authorized_for_configuration_information=nagiosadmin,admin
# 设置可以关闭或重启 Nagios 的账户名称
```

```
authorized_for_system_commands=nagiosadmin,admin
#设置可以查看所有服务信息的账户名称
authorized_for_all_services=nagiosadmin,admin
#设置可以查看所有主机信息的账户名称
authorized_for_all_hosts=nagiosadmin,admin
#设置允许执行服务相关命令的账户名称
authorized_for_all_service_commands=nagiosadmin,admin
#设置允许执行主机相关命令的账户名称
authorized_for_all_host_commands=nagiosadmin,admin
```

（3）修改命令配置文件 commands.cfg，该文件定义具体的命令实现方式。如发送报警邮件具体使用什么工具、邮件等内容的格式定义，在末尾手动添加此内容。

```
[root@nagios ~]# vi /usr/local/nagios/etc/objects/commands.cfg
……
define command{
        command_name    check_nrpe
        command_line    $USER1$/check_nrpe -H $HOSTADDRESS$ -c $ARG1$
}
```

（4）contacts.cfg 文件可以根据需求设置需要联系的联系人，在发生故障之后，就会调用该文件中的联系人和联系方式。

```
[root@nagios ~]# vi /usr/local/nagios/etc/objects/contacts.cfg
```

localhost.cfg 文件为本机配置文件，用于设置如何监控本机服务器资源。

templates.cfg 文件为模板配置文件，主要定义一些其他配置文件需要调用的定义，包含报警方式、报警时间、报警选项等。

（5）修改 nrpe 插件配置文件，在服务中创建用于监控远程主机所需要的命令。

```
[root@nagios ~]# vi /usr/local/nagios/etc/nrpe.cfg
……
修改
allowed_hosts=127.0.0.1,192.168.200.10
添加
command[check_disk]=/usr/local/nagios/libexec/check_disk -w 20% -c 10%
……
```

（6）监控的远程主机为 web2 和 web3。我们需要在相应目录下创建两个 web2.cfg 和 web3.cfg 文件，可使用 localhost.cfg 作为参考模板。以下为 web2.cfg 全文，此处以 web2 节点为例，web3 只需要更改主机名称、IP 地址、主机组名称即可。

```
[root@nagios ~]# cd /usr/local/nagios/etc/objects/
[root@nagios objects]# cp localhost.cfg web2.cfg
[root@nagios objects]# vi web2.cfg
define host{
        use                     linux-server
        host_name               web2
        alias                   web2.example.com
        address                 192.168.200.20
        }
define hostgroup{
        hostgroup_name          webs
        alias                   Linux Servers
        members                 web2
        }
```

```
define service{
    use                     generic-service
    host_name               web2
    service_description     PING
    check_command           check_ping!100.0,20%!500.0,60%
    notifications_enabled   1
    }
define service{
    use                     generic-service
    host_name               web2
    service_description     Sys_load
    check_command           check_nrpe!check_load
    notifications_enabled   1
    }
define service{
    use                     generic-service
    host_name               web2
    service_description     Current Users
    check_command           check_nrpe!check_users
    notifications_enabled   1
    }
define service{
    use                     generic-service
    host_name               web2
    service_description     Total Processes
    check_command           check_nrpe!check_total_procs
    notifications_enabled   1
    }
define service{
    use                     generic-service
    host_name               web2
    service_description     SSH
    check_command           check_ssh
    notifications_enabled   1
    }
define service{
    use                     generic-service
    host_name               web2
    service_description     HTTP
    check_command           check_http
    notifications_enabled   1
    }
```

（7）全部配置完成之后，在 nagios 节点重新加载所有配置文件，命令和结果如下所示：

```
[root@nagios ~]# /usr/local/nagios/bin/nrpe -c /usr/local/nagios/etc/nrpe.cfg -d
[root@nagios ~]# /etc/init.d/nagios restart
Restarting nagios (via systemctl):                          [  OK  ]
```

3. 客户端安装与配置

（1）将下面这些操作分别应用于 web2 和 web3 节点，安装依赖包并创建账户和组，此处以 web2 为例，命令如下所示：

```
[root@web2 nagios-plugins-2.3.3]# yum -y install gcc glibc glibc-common make net-snmp openssl openssl-devel httpd
[root@web2 ~]# groupadd nagios
```

```
[root@web2 ~]# useradd -g nagios -s /sbin/nologin nagios
```

（2）将 nagios-plugin 和 nrpe 软件压缩包上传至 web2 和 web3 节点，使用命令进行解压，然后进行编译安装，此处以 web2 为例，命令如下所示：

```
[root@web2 ~]# tar -zxvf nagios-plugins-2.3.3.tar.gz -C /usr/src/
[root@web2 ~]# cd /usr/src/nagios-plugins-2.3.3
[root@web2 nagios-plugins-2.3.3]# ./configure --with-nagios-user=nagios --with-nagios-group=nagios --prefix=/usr/local/nagios
[root@web2 nagios-plugins-2.3.3]# make
[root@web2 nagios-plugins-2.3.3]# make install
[root@web2 nagios-plugins-2.3.3]# chown -R nagios.nagios /usr/local/nagios/
[root@web2 ~]# tar -zxvf nrpe-2.14.tar.gz -C /usr/src/
[root@web2 ~]# cd /usr/src/nrpe-2.14/
[root@web2 nrpe-2.14]# ./configure
[root@web2 nrpe-2.14]# make all
[root@web2 nrpe-2.14]# make install-plugin
[root@web2 nrpe-2.14]# make install-daemon
[root@web2 nrpe-2.14]# make install-daemon-config
```

（3）编译安装完成之后，在 web2 和 web3 节点修改 NRPE 配置文件，此处以 web2 为例，命令如下所示：

```
[root@web2 ~]# vi /usr/local/nagios/etc/nrpe.cfg
……
修改
allowed_hosts=127.0.0.1,192.168.200.10
添加
command[check_disk]=/usr/local/nagios/libexec/check_disk -w 20% -c 10%
……
```

（4）全部配置完成之后，在 web2 节点重新加载所有配置文件，并将 nrpe 设置为开机自启，使用命令查看是否有 5666 端口，此处以 web2 为例，命令和结果如下所示：

```
[root@web2 ~]# /usr/local/nagios/bin/nrpe -c /usr/local/nagios/etc/nrpe.cfg -d
[root@web2 ~]# echo "/usr/local/nagios/bin/nrpe -c /usr/local/nagios/etc/nrpe.cfg -d" >> /etc/rc.local
[root@web2 ~]# netstat -nutpl | grep 5666
[root@web2 ~]# systemctl restart httpd
```

4. 服务测试

（1）登录 nagios 节点，通过 check_nrpe 命令检测被监控端相关的性能参数，单独使用 check_nrpe 命令可以检测被监控端的 NRPE 版本号，命令和结果如下所示：

```
[root@nagios ~]# /usr/local/nagios/libexec/check_nrpe -H 192.168.200.10
NRPE v2.14
[root@nagios ~]# /usr/local/nagios/libexec/check_nrpe -H 192.168.200.20
NRPE v2.14
[root@nagios ~]# /usr/local/nagios/libexec/check_nrpe -H 192.168.200.30
NRPE v2.14
```

（2）通过客户端浏览器访问 http://192.168.200.10/nagios，首先会弹出图 7-1 所示的访问登录窗口，输入之前设置的用户名和密码即可登录（admin：000000）。

（3）客户端访问监控服务器的 Web 服务效果如图 7-2、图 7-3 所示。

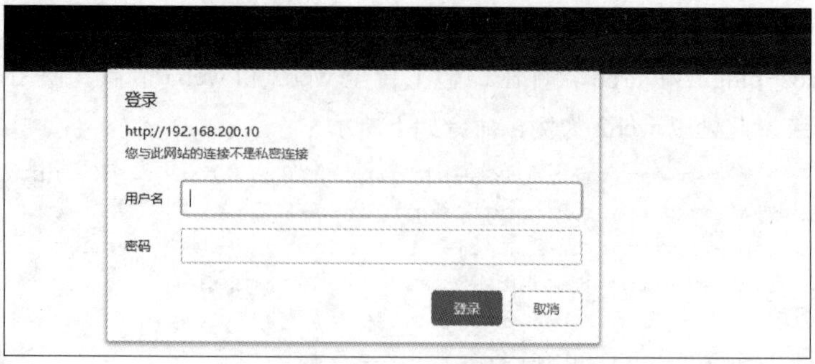

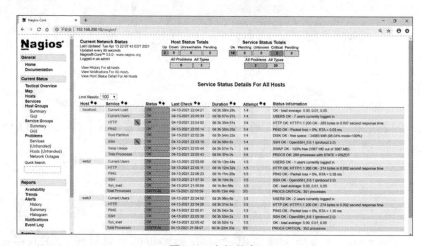

图 7-1　窗口验证

图 7-2　主机信息

Host	Status	Last Check	Duration	Status Information
localhost	UP	04-13-2021 23:37:05	0d 5h 28m 6s	PING OK - Packet loss = 0%, RTA = 0.04 ms
web2	UP	04-13-2021 23:32:45	0d 2h 43m 22s	PING OK - Packet loss = 0%, RTA = 1.29 ms
web3	UP	04-13-2021 23:36:15	0d 5h 27m 16s	PING OK - Packet loss = 0%, RTA = 1.24 ms

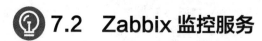

图 7-3　主机列表

本任务到此结束。

7.2　Zabbix 监控服务

学习目标

学完本节后，您应能够：

- 了解 Zabbix 服务原理。
- 掌握 Zabbix 服务安装与运维。
- 掌握 Zabbix 监控案例实施。

7.2.1　Zabbix 服务和原理

1. 什么是 Zabbix

Zabbix 是一个企业级分布式开源监控解决方案。

Zabbix 支持主动轮询（polling）和被动捕获（trapping）。Zabbix 所有的报表、统计数据和配置参数都可以通过基于 Web 的前端页面进行访问。基于 Web 的前端页面确保您可以在任何地方访问您监控的网络状态和服务器健康状况。适当的配置后，Zabbix 可以在监控 IT 基础设施方面发挥重要作用。无论是对于有少量服务器的小型组织，还是拥有大量服务器的大企业而言，同样适用。

Zabbix 是免费的。Zabbix 是根据 GPL 通用公共许可证的第二版编写和发布的。这意味着产品源代码是免费发布的，可供公共使用。

Zabbix 是一个用于网络，操作系统和应用程序的开源监控软件，由拉脱维亚的 Alexei Vladishev 创建。它旨在监视和跟踪各种网络服务，服务器和其他网络硬件的状态。

Zabbix 可以使用 MySQL、MariaDB、PostgreSQL、SQLite、Oracle 或 IBM DB2 来存储数据。它的后端用 C 语言编写，Web 前端用 PHP 编写。Zabbix 提供多种监控选项：

（1）简单检查可以验证标准服务（如 SMTP 或 HTTP）的可用性和响应性，而无须在受监视主机上安装任何软件。

（2）可以在 UNIX 和 Windows 主机上安装 Zabbix 代理，以监视 CPU 负载，网络利用率，磁盘空间等统计信息。

（3）作为在主机上安装代理的替代方法，Zabbix 支持通过 SNMP、TCP 和 ICMP 检查以及 IPMI、JMX、SSH、Telnet 和使用自定义参数进行监控。Zabbix 支持各种近实时通知机制，包括 XMPP。

Zabbix 由两部分构成：zabbix server 与可选组件 zabbix agent。

2. Zabbix 架构及组件

图 7-4 所示为 Zabbix 服务的大致架构，Zabbix 监控组件主要包括：Zabbix Server、Zabbix Proxy、Zabbix Agent；其中 Zabbix Server 服务包括：WEB GUI、Database、Zabbix Server。而且 Zabbix Web GUI 和 Zabbix Databases 及 Zabbix Server 不一定在同一主机上，也可以是分开的。

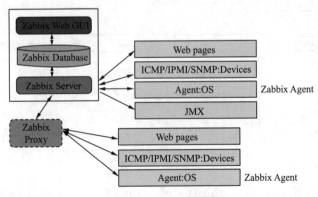

图 7-4　Zabbix 服务架构

组件功能介绍：

（1）Zabbix Agent：部署在被监控主机上，负责被监控主机的数据，并将数据发送给 Zabbix Server 或 Zabbix Proxy。

（2）Zabbix Server：负责接收 Agent 发送的报告信息，并且负责组织配置信息、统计信息、操作数据等。

（3）Zabbix Database：用于存储所有 Zabbix 的配置信息、监控数据的数据库。

（4）Zabbix Web：Zabbix 的 Web 界面，管理员通过 Web 界面管理 Zabbix 配置以及查看

Zabbix 相关监控信息,可以单独部署在独立的服务器上。

(5)Zabbix Proxy:可选组件,用于分布式监控环境中,Zabbix Proxy 代表 Server 端,完成局部区域内的信息收集,最终统一发往 Server 端。

3. Zabbix 主要特点

(1)安装与配置简单,学习成本低。

(2)支持多语言(包括中文)。

(3)免费开源。

(4)自动发现服务器与网络设备。

(5)分布式监视以及 Web 集中管理功能。

(6)可以无 Agent 监视。

(7)用户安全认证和柔软的授权方式。

(8)通过 Web 界面设置或查看监视结果。

(9)E-mail 等通知功能。

4. Zabbix 监控流程

图 7-5 所示为 Zabbix 监控系统具体流程图。

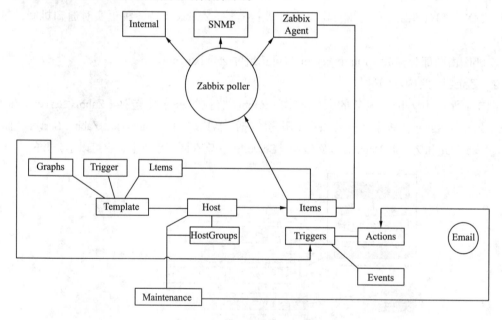

图 7-5 Zabbix 监控流程图

Agentd 安装在被监控的主机上,Agent 负责定期收集客户端本地各项数据,并发送至 Zabbix Server 端,Zabbix Server 收到数据,将数据存储到数据库中,用户基于 Zabbix Web 可以看到数据在前端展现图像。当 Zabbix 监控某个具体的项目,改项目会设置一个触发器阈值,当被监控的指标超过该触发器设定的阈值,会进行一些必要的动作,动作包括:发送信息(邮件、微信、短信)、发送命令(Shell 命令、Reboot、Restart、Install 等)。

Zabbix 监控部署在系统中,包含常见的 5 个程序:zabbix_server、zabbix_agentd、zabbix_proxy、zabbix_get、zabbix_sender 等。

7.2.2　Zabbix 服务安装与运维

1. 安装服务

（1）安装最小化 CentOS 7.2-1511 操作系统，配置节点的主机名与 IP 地址，并使用远程终端工具进行连接，命令如下所示：

```
[root@localhost ~]# hostnamectl set-hostname zabbix-server
[root@localhost ~]# logout
[root@zabbix-server ~]#
[root@localhost ~]# hostnamectl set-hostname zabbix-agent
[root@localhost ~]# logout
[root@zabbix-agent ~]#
```

（2）安装所需要的服务，包括数据库和 zabbix 服务包，命令和结果如下所示：

```
[root@zabbix-server ~]# yum install httpd mariadb-server mariadb -y
[root@zabbix-server ~]# yum install zabbix-server-mysql zabbix-web-mysql zabbix-agent -y
```

升级 trousers 服务。

```
[root@zabbix-server ~]#  yum install trousers -y
```

（3）使用命令启动 httpd 服务，并设置开机自启，命令如下所示：

```
[root@zabbix-server ~]# systemctl start httpd
[root@zabbix-server ~]# systemctl enable httpd
Created symlink from /etc/systemd/system/multi-user.target.wants/httpd.service to /usr/lib/systemd/system/httpd.service.
```

（4）启动数据库服务，并设置开机自启，登录到数据库，创建 zabbix 库授予 zabbix 用户的访问权限，命令和结果如下所示：

```
[root@zabbix-server ~]# systemctl start mariadb
[root@zabbix-server ~]#  systemctl enable mariadb
Created symlink from /etc/systemd/system/multi-user.target.wants/mariadb.service to /usr/lib/systemd/system/mariadb.service.
[root@zabbix-server ~]# mysql
Welcome to the MariaDB monitor.  Commands end with ; or \g.
Your MariaDB connection id is 2
Server version: 5.5.68-MariaDB MariaDB Server
Copyright (c) 2000, 2018, Oracle, MariaDB Corporation Ab and others.
Type 'help;' or '\h' for help. Type '\c' to clear the current input statement.
MariaDB [(none)]> create database zabbix character set utf8 collate utf8_bin;
Query OK, 1 row affected (0.00 sec)
MariaDB [(none)]> grant all privileges on zabbix.* to zabbix@'%' identified by 'zabbix';
Query OK, 0 rows affected (0.00 sec)
MariaDB [(none)]> grant all privileges on zabbix.* to zabbix@localhost identified by 'zabbix';
Query OK, 0 rows affected (0.00 sec)
```

（5）退出数据库之后，进入 /usr/share/doc/zabbix-server-mysql-4.0.30 目录，使用命令导入数据库文件，命令如下所示：

```
[root@zabbix-server ~]# cd /usr/share/doc/zabbix-server-mysql-4.0.30/
[root@zabbix-server zabbix-server-mysql-4.0.30]# ll
total 2312
-rw-r--r--. 1 root root       98 Mar 29 04:33 AUTHORS
```

```
-rw-r--r--. 1 root root 1014545 Mar 29 04:33 ChangeLog
-rw-r--r--. 1 root root   17990 Mar 29 04:33 COPYING
-rw-r--r--. 1 root root 1316241 Mar 29 05:07 create.sql.gz
-rw-r--r--. 1 root root      52 Mar 29 04:33 NEWS
-rw-r--r--. 1 root root    1317 Mar 29 04:33 README
[root@zabbix-server zabbix-server-mysql-4.0.30]# zcat create.sql.gz | mysql -uroot zabbix
```

2．配置服务

（1）编辑 /etc/php.ini 文件，设置时区，在 [Date] 字段下设置 date.timezone=PRC，配置如下：

```
[root@zabbix-server ~]# vi /etc/php.ini
[Date]
; Defines the default timezone used by the date functions
; http://php.net/date.timezone
;date.timezone = PRC
```

（2）编辑 /etc/httpd/conf.d/zabbix.conf 文件，修改时区，修改 php_value date.timezone 为 Asia/Shanghai，然后重启服务，配置如下：

```
[root@zabbix-server ~]# vi /etc/httpd/conf.d/zabbix.conf
…
php_value date.timezone Asia/Shanghai
[root@zabbix-server ~]# systemctl restart httpd
```

（3）修改 Zabbix 配置文件并启动，修改 /etc/zabbix/zabbix_server.conf 配置文件，修改完的配置文件如下所示：

```
[root@zabbix-server ~]# vi /etc/zabbix/zabbix_server.conf
DBHost=localhost
DBPassword=zabbix
DBSocket=/var/lib/mysql/mysql.sock
```

（4）启动 zabbix 服务，查看端口号，验证 zabbix-server 的服务端口 10051 是否存在，命令和结果如下所示：

```
[root@zabbix-server ~]#  systemctl start zabbix-server
[root@zabbix-server ~]# netstat -ntpl
Active Internet connections (only servers)
Proto Recv-Q Send-Q Local Address      Foreign Address    State       PID/Program name
tcp        0      0 0.0.0.0:3306       0.0.0.0:*          LISTEN      10647/mysqld
tcp        0      0 0.0.0.0:22         0.0.0.0:*          LISTEN      1419/sshd
tcp        0      0 127.0.0.1:25       0.0.0.0:*          LISTEN      1734/master
tcp        0      0 0.0.0.0:10051      0.0.0.0:*          LISTEN      10763/zabbix_server
tcp6       0      0 :::80              :::*               LISTEN      10722/httpd
tcp6       0      0 :::22              :::*               LISTEN      1419/sshd
tcp6       0      0 ::1:25             :::*               LISTEN      1734/master
tcp6       0      0 :::10051           :::*               LISTEN      10763/zabbix_server
```

3．使用服务

（1）安装完成，在浏览器中访问 http://192.168.200.10/zabbix，进入 Zabbix 安装向导，单击右下角 "Next step" 按钮，进入下一步操作，如图 7-6 所示。

第 7 章　监控服务与自动化运维工具

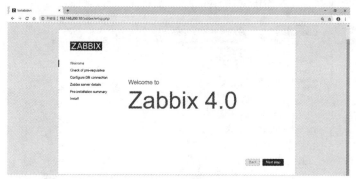

图 7-6　安装界面

（2）显示 PHP 版本信息等内容，然后单击右下角"Next step"按钮，进入下一步操作，如图 7-7 所示。

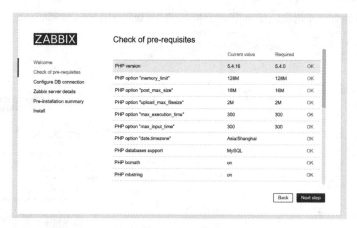

图 7-7　状态信息

（3）填写连接数据库的必要信息，Password 为 zabbix，填写内容如图 7-8 所示，填写完毕后单击右下角"Next step"按钮，进行下一步操作。

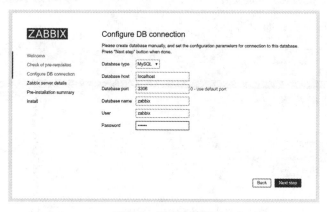

图 7-8　数据库信息

（4）填写 Zabbix 服务端的详细信息，内容如图 7-9 所示，Name 字段可以随意填写，给监控平台命名。然后单击右下角"Next step"按钮，进行下一步操作。

图 7-9　服务端信息

（5）安装 Zabbix，安装成功后，单击右下角"Finish"按钮，结束安装，如图 7-10 所示。

（6）单击"Finish"按钮后，进入登录界面，使用默认的用户名和密码 Admin/zabbix 登录，如图 7-11 所示。

图 7-10　完成安装

图 7-11　Web 监控界面登录

（7）进入 Zabbix 主页，如图 7-12 所示。

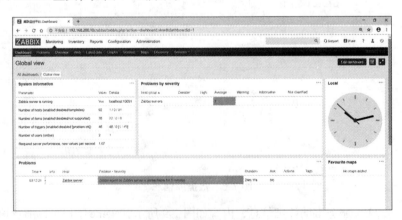

图 7-12　监控主页界面

（8）单击右上角的头像按钮，进入设置界面。将"Language"一栏修改为"Chinexe(zh-CN)"，然后单击"Update"按钮，如图 7-13 所示。

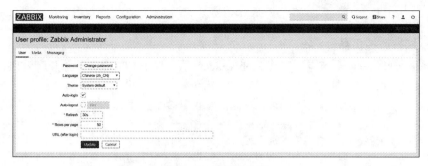

图 7-13　设置语言

（9）中文界面的 Zabbix 监控界面已配置完成，如图 7-14 所示。

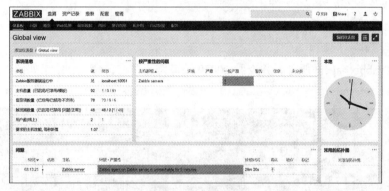

图 7-14　设置完成

7.2.3　任务：部署 Zabbix 分布式监控系统

Zabbix 软件能够监控众多网络参数和服务器的健康度、完整性。Zabbix 使用灵活的告警机制，允许用户为几乎任何事件配置基于邮件的告警。这样用户可以快速响应服务器问题。Zabbix 基于存储的数据提供出色的报表和数据可视化功能。

任务执行清单

在本任务中，您将通过实践部署 Zabbix 分布式监控系统。

目标

- 掌握监控机器的添加。
- 掌握 Zabbix 监控 Apache 服务。

重要信息

- 本任务采用 CentOS 7 操作系统。
- 虚拟机密码为 000000。
- 提前配置好 yum 源仓库。
- 关闭防火墙和 selinux。
- 提前安装好 zabbix-server。

解决方案

1. 添加被监控机器

（1）登录 zabbix-agent 节点，安装 zabbix-agent 服务和 HTTP 服务，并开启 HTTP 服务设置开机启动。命令如下：

```
[root@zabbix-agent ~]# yum install -y zabbix-agent httpd
[root@zabbix-agent ~]# systemctl start httpd
[root@zabbix-agent ~]# systemctl enable httpd
Created symlink from /etc/systemd/system/multi-user.target.wants/httpd.service to /usr/lib/systemd/system/httpd.service.
```

（2）修改 /etc/zabbix/zabbix_agentd.conf 配置文件，配置 zabbix-agent，修改如下所示：

```
[root@zabbix-agent ~]# vi /etc/zabbix/zabbix_agentd.conf
Server=192.168.200.10
ServerActive=192.168.200.10
Hostname=Zabbix-agent
```

（3）启动 zabbix-agent 服务，并查看 10050 端口是否存在，命令和结果如图 7-15 所示。

```
[root@zabbix-agent ~]# systemctl start zabbix-agent
```

图 7-15　端口信息

2. 添加主机

（1）回到 Web 界面，选择菜单栏"配置→主机→创建主机"命令，会跳转到图 7-16 所示的配置主机的页面，在界面中添加如下信息，然后选择"模板"选项卡。

图 7-16　添加主机

（2）在模板设置中填写信息，将需要使用的模板链接添加至该选项中，如图 7-17 所示。单击"添加"按钮，选择"添加"用于创建主机，zabbix-agent 节点被添加到监控中，如图 7-18 所示。

第 7 章 监控服务与自动化运维工具

图 7-17 添加模板

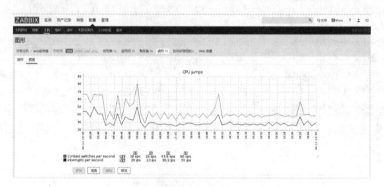

图 7-18 主机添加成功

选择图形按钮，选择需要查看的图形进行预览，此处选择为 CPU jumps，如图 7-19 所示。

3. 添加监控 Apache 服务

（1）在 zabbix-agent 节点中下载脚本到 Linux 主机，然后安装 httpd 服务，命令如下所示：

```
[root@zabbix-agent ~]# wget https://github.com/lorf/zapache/archive/master.zip
[root@zabbix-agent ~]# yum -y install httpd
```

图 7-19 主机监控图

（2）进入 Apache 服务配置文件中，在最后一行添加此文件，然后重启服务，命令和结果如下所示：

```
[root@zabbix-agent ~]# cd /etc/httpd/
[root@zabbix-agent httpd]# vi conf/httpd.conf
ExtendedStatus On
<location /server-status>
   SetHandler server-status
   Order allow,deny
   Allow from 127.0.0.1 192.168.200.0/24
</location>
[root@zabbix-agent2 httpd]# systemctl restart httpd
```

（3）使用浏览器输入 192.168.200.20/server-status，查看服务状态是否启动成功，结果如图 7-20 所示。

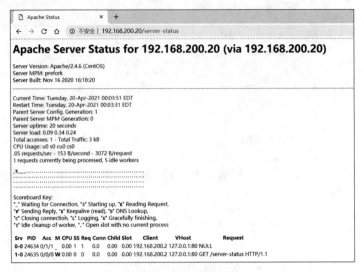

图 7-20　服务网页测试

（4）解压压缩包，将配置文件复制到 zabbix 配置目录下，然后更改文件名称和修改配置文件信息，命令和结果如下所示：

```
[root@zabbix-agent ~]# unzip master.zip
[root@zabbix-agent ~]# cd zapache-master/
[root@zabbix-agent zapache-master]# cp zapache /usr/local/bin/
[root@zabbix-agent zapache-master]# cp userparameter_zapache.conf.sample /etc/zabbix/zabbix_agentd.d/
[root@zabbix-agent zapache-master]# cd /etc/zabbix/zabbix_agentd.d/
[root@zabbix-agent zabbix_agentd.d]# mv userparameter_zapache.conf.sample userparameter_zapache.conf
[root@zabbix-agent zabbix_agentd.d]# vi userparameter_zapache.conf
UserParameter=zapache[*],/usr/local/bin/zapache \$1
```

（5）配置完成之后，登录 zabbix Web 界面，在模板选项中选择"导入"，加载导入的文件，界面如图 7-21 所示。

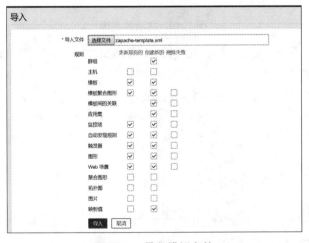

图 7-21　导入模板文件

（6）在创建好的主机"web 服务器"中，更新刚刚添加的 Apache 模板文件，如图 7-22 所示。

第 7 章 监控服务与自动化运维工具

图 7-22 添加模板到主机

（7）在 Web 平台中选择"最新数据"选项卡，筛选 Apache，在设置完成后，勾选想要显示的最新数据图形，如图 7-23 所示，单击 Web 服务器下方的"图形"链接即可查看监控数据图形。

图 7-23 查看最新数据图形

本任务到此结束。

7.3 Ansible 自动化运维

学习目标

学完本节后，您应能够：
- 了解 Ansible 服务原理。
- 了解 Ansible 任务执行。
- 掌握 Ansible 自动化案例实施。

7.3.1 Ansible 服务

1. Ansible 简介

Ansible 是一种自动化运维工具，它是一种集成 IT 系统的配置管理、应用部署、执行特定任务的开源平台，它是基于 Anon 的语言，由 Paramiko 和 PyYAML 两个关键模块构建。Ansible 集合了众多运维工具（Puppet、Cfengine、Chef、Func、Fabric）的优点，实现了批量系统配置、批量程序部署、批量运行命令等功能，默认通过 SSH 协议管理机器。

2. Ansible 架构

Ansible 是基于模块工作的，本身没有批量部署的能力，真正具有批量部署能力的是 Ansible 所运行的模块，Ansible 只是提供了运行一种框架，如图 7-24 所示。

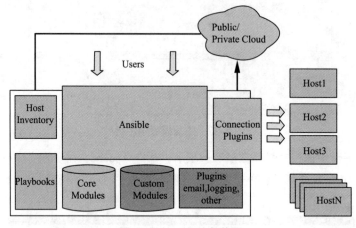

图 7-24　Ansible 架构图

3. Ansible 模块介绍

（1）Ansible：Ansible 核心程序。

（2）Playbooks：任务剧本（任务集），编排定义 Ansible 任务集的配置文件，由 Ansible 依次执行多个任务，通常是 JSON 格式的 YML 文件。

（3）Host Inventory：Ansible 管理主机的清单，记录由 Ansible 管理的主机信息，包括端口、密码、IP 等。

（4）Modules：核心模块，主要操作是通过调用核心模块来完成管理任务。

（5）Custom Modules：自定义模块，完成核心模块无法完成的功能，支持多种语言。

（6）Connection Plugins：基于连接插件连接到各个主机上，即 Ansible 和 Host 通信使用，默认是使用 SSH。

（7）Plugins：模块功能的补充，如连接类型插件、循环插件、变量插件等，可借助于插件完成更丰富的功能。

（8）Playbooks：批量的命令文件。

用户请求发送给 Ansible 核心模块，Ansible 核心模块通过 Host inventory 模块寻找需要运行的主机，然后通过 Connection plugins 连接远程的主机，并发送命令。Ansible 使用插件来连接每一个被控制端 Host，此外，也通过插件来记录日志等信息。

7.3.2　Ansible 任务执行

1. Ansible 任务执行模式

Ansible 任务执行分为以下两种模式：

（1）ad-hoc 模式。使用单个模块，一次执行多条命令，相当于在 bash 中执行一句 Shell 命令。

（2）playbook 模式。Ansible 主要的管理方式。通过多个 task 的集合完成一类功能，可以理解为多个 ad-hoc 模式的合成。

2. Ansible 执行流程

Ansible 执行流程如图 7-25 所示。

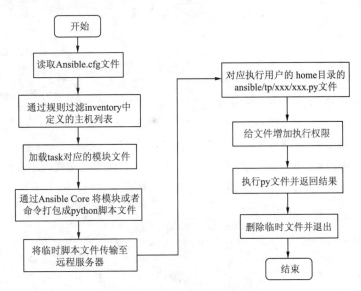

图 7-25　Ansible 执行流程图

7.3.3　任务：Ansible 部署 Nginx 服务

下面看一个使用 Playbook 部署服务的例子。在这个例子中，我们将使用 Ansible 配置一台服务器运行 nginx 进程。

任务执行清单

在本任务中，您将通过实践部署 Nginx 服务。

目标

- 掌握 Ansible 安装与部署。
- 掌握 Ansible 一键部署 Nginx。

重要信息

- 本任务采用 CentOS 7 操作系统。
- 虚拟机密码为 000000。
- 提前配置好 yum 源仓库。
- 关闭防火墙和 selinux。

解决方案

1．基础环境配置

（1）登录到虚拟机，更改虚拟机的主机名，命令和结果如下所示：

```
[root@localhost ~]# hostnamectl set-hostname ansible
[root@localhost ~]# logout
[root@ansible ~]#
```

（2）在 Ansible 服务端运行命令，ping 模块检查网络连通性，command 模块执行 Shell 命令，command: 作为 Ansible 的默认模块，可以运行远程权限范围内的所有 Shell 命令。

```
[root@ansible ~]# ansible -i /etc/ansible/hosts webserver -m ping
192.168.200.10 | SUCCESS => {
    "ansible_facts": {
        "discovered_interpreter_python": "/usr/bin/python"
    },
    "changed": false,
    "ping": "pong"
}
```

(3)在虚拟机中使用 epel 源安装 ansible 服务,命令和结果如下所示:

```
[root@ansible ~]# yum install epel-release -y
[root@ansible ~]# yum install -y ansible
```

(4)基于 ssh 密钥来访问、定义主机清单,一般来说,使用明文密码不安全,所以增加主机无密码访问。在 Ansible 服务端生成密钥,并且复制公钥到节点中。

```
[root@ansible ~]# ssh-keygen
Generating public/private rsa key pair.
Enter file in which to save the key (/root/.ssh/id_rsa):
Created directory '/root/.ssh'.
Enter passphrase (empty for no passphrase):
Enter same passphrase again:
Your identification has been saved in /root/.ssh/id_rsa.
Your public key has been saved in /root/.ssh/id_rsa.pub.
The key fingerprint is:
72:70:e0:73:20:5d:19:0a:23:3a:63:2c:4c:fc:18:b8 root@ansible
The key's randomart image is:
+--[ RSA 2048]----+
|o.. +.o.oo       |
|=+ . =.+.        |
|==+   = o        |
|E+ .   =         |
|     . S         |
|      o          |
|                 |
|                 |
|                 |
+-----------------+
[root@ansible ~]# ssh-copy-id  root@192.168.200.10
[root@ansible ~]# ssh-copy-id  root@192.168.200.20
```

2. 服务目录编写

(1)使用命令初始化一个 role 角色,便于后期使用,然后查看所创建的角色。命令和结果如下所示:

```
[root@ansible ~]#  ansible-galaxy init /etc/ansible/roles/webserver
- Role /etc/ansible/roles/webserver was created successfully
[root@ansible ~]# ls /etc/ansible/roles/
webserver
```

把初始化后 role 里面没用的目录删除,没有的目录就创建,按照第一步的目录架构来进行。

(2)配置 ansible.cfg 的配置文件,用于定义用户、端口、文件路径等信息。

```
[root@ansible ansible]# vi ansible.cfg
[defaults]
```

```
inventory = /etc/ansible/hosts
sudo_user=root
remote_port=22
host_key_checking=False
remote_user=root
log_path=/var/log/ansible.log
module_name=command
private_key_file=/root/.ssh/id_rsa
```

（3）group_vars 主要用于主机组变量，目录里包含以组名命名的 yaml 文件，创建一个用户组，在文件夹中配置变量 all 文件，此处的文件名称只能写 all，命令和结果如下所示：

```
[root@ansible ansible]# mkdir group_vars
[root@ansible ansible]# vi group_vars/all
---
# vars file for /etc/ansible/roles/webservs
worker_processes: 4
worker_connections: 768
max_open_files: 65506
```

（4）Roles 角色信息。

角色目录中通常存放着 tasks、handler、vars、templates、meta 子目录，这些子目录并不是必需的，如果没用到某个目录（如没有用到模板 templates），可以为空目录，或者不创建。基于这样的目录结构，role 会自动加载到目录内的 tasks、vars 及 handlers。

在 /etc/ansible 目录下，配置 site.yaml 文件作为执行入口文件，文件中一般用于定义所使用的 roles 操作，结果如下所示：

```
[root@ansible ansible]# vi site.yaml
---
# this playbook deploy the whole application stack in this site
- name: configuration and deploy webserver and application code
  hosts: webserver
  roles:
    - webserver
```

（5）在角色目录下，定义 handlers 文件的信息，也就是我们使用的触发器，用于部署 Nginx 服务中的服务启动。

```
[root@ansible ~]# cd /etc/ansible/roles/webserver/
[root@ansible webserver]# vi handlers/main.yml
---
# handlers file for /etc/ansible/roles/webserver
- name: restart nginx
  service: name=nginx state=restarted
```

（6）play 中运行的任务命令，也就是执行的哪些 ansible 模块，如 command、shell、service、yum 等。配置 tasks 文件，用于存放执行操作的 yaml 文件信息。

```
[root@ansible webserver]# vi tasks/main.yml
---
# tasks file for /etc/ansible/roles/webserver
- include: install_nginx.yaml
[root@ansible webserver]# vi tasks/install_nginx.yaml
---
# tasks file for /etc/ansible/roles/webserver
```

```yaml
    - name: yum nginx
      yum:
        name:
          - nginx
        state: present
    - name: copy index.html
      template:
        src: index.html.j2
        dest: /usr/share/nginx/html/index.html
        mode: 0644
      notify: restart nginx
    - name: see file
      service:
        name: nginx
        state: started
```

（7）使用 template 模板渲染功能时，所需的模块文件存放在这个目录，编写一个网页显示文件，如下所示：

```
[root@ansible templates]# vi index.html.j2
<html>
  <head>
    <title>welcome to 192.168.200.10</title>
  </head>
  <body>
  <h1>192.168.200.10-nginx, confitured by ansible</h1>
  <p>if you can see this, ansible successfully installed nginx.</p>

  <p>{{ ansible_hostname }}</p>
  </body>
</html>
```

3. 服务部署

（1）使用如下命令执行部署，结果如图 7-26 所示。

```
[root@ansible ansible]# ansible-playbook site.yaml
```

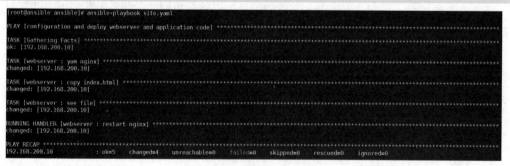

图 7-26　服务部署成功

（2）ansible 部署完成之后，打开浏览器，输入服务端 IP：192.168.200.10，即可看到图 7-27 所示网页结果。

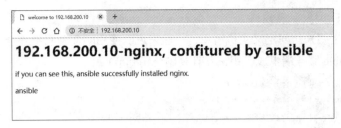

图 7-27　网页测试

本任务到此结束。

7.4　开放研究任务：Ansible 批量部署 LAMP 环境

下面我们将以 Ansible 应用案例，即批量部署 LAMP 应用的例子，来学习 Ansible 的最佳部署方式。

任务执行清单

在本任务中，将通过安装和部署 Ansible 服务，使用 playbook 批量部署 LAMP 环境。

（1）基础环境配置。

（2）构建 Ansible 服务。

（3）服务部署与测试。

目标

- 掌握 Ansible 服务的安装。
- 掌握 Ansible 文件的配置。
- 掌握 Ansible 批量部署 LAMP。

重要信息

- 本任务采用 CentOS 7 操作系统。
- 虚拟机密码为 000000。
- 提前配置好 yum 源仓库。
- 关闭防火墙和 Selinux。

解决方案

任务执行清单

在本任务中，您将通过实验安装和部署 Ansible 服务，使用 playbook 批量部署 LAMP 环境。

1. 基础环境配置

（1）创建两台虚拟机 192.168.200.10（ansible）、192.168.200.20（client），登录到虚拟机，更改虚拟机的主机名，命令和结果如下所示：

```
[root@localhost ~]# hostnamectl set-hostname ansible
[root@localhost ~]# logout
[root@ansible ~]#
[root@localhost ~]# hostnamectl set-hostname client
[root@localhost ~]# logout
```

```
[root@client ~]#
```

（2）基于 ssh 密钥来访问定义主机清单，一般来说，使用明文密码不安全，所以增加主机无密码访问。在 Ansible 服务端生成密钥，并且复制公钥到节点中。

```
[root@ansible ~]# ssh-keygen
[root@ansible ~]# ssh-copy-id root@192.168.200.10
[root@ansible ~]# ssh-copy-id root@192.168.200.20
```

（3）配置好 yum 源之后，在 ansible 节点安装 httpd 服务，命令如下所示：

```
[root@ansible ~]#  yum install httpd -y
```

（4）安装并配置数据库服务，重新定义数据库数据存储的位置，并启动数据库服务，命令和结果如下所示：

```
[root@ansible ~]# yum install mariadb-server  mariadb  -y
[root@ansible ~]# mkdir -p /mydata/data
[root@ansible ~]# chown -R mysql:mysql /mydata/
[root@ansible ~]# vi /etc/my.cnf
datadir=/mydata/data
[root@ansible ~]# systemctl start mariadb
```

（5）安装 PHP 服务和 php-mysql 模块，创建一个测试 PHP 服务的网页文件，并在浏览器中输入 ansible 节点 IP 进行网页查看，如图 7-28 所示。

```
[root@ansible ~]# yum install php php-mysql -y
[root@ansible ~]# vi /var/www/html/index.php
<?php
    phpinfo();
?>
[root@ansible ~]# systemctl restart httpd
```

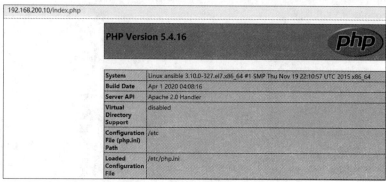

图 7-28　网页测试

2．构建 Ansible 服务

（1）在 ansible 节点安装 ansible 服务，并在配置文件中定义 web-servers 组名和 IP 信息，命令如下所示。

```
[root@ansible ~]# yum -y install ansible
[root@ansible ~]# vi /etc/ansible/hosts
[web-servers]
192.168.200.10
192.168.200.20
```

（2）使用 playbook 创建一个 LAMP 构建的任务，首先创建相关的目录及文件。将上面搭建成功的 LAMP 环境的 httpd 和 MySQL 的配置文件复制到对应目录下，命令如下所示：

```
[root@ansible ~]# mkdir -pv
/etc/ansible/lamp/roles/{prepare,httpd,mysql,php}/{tasks,files,templates,vars,meta,default,handlers}
[root@ansible ~]# cd /etc/ansible/
[root@ansible ansible]# cp /etc/httpd/conf/httpd.conf  lamp/roles/httpd/files/
[root@ansible ansible]# cp /etc/my.cnf   lamp/roles/mysql/files/
```

（3）在 roles 角色目录中，存在一个 prepare 角色信息，其中包含重要的配置文件，下面需要创建一个 prepare 的 playbooks（在此步骤之前，两台主机都需安装 wget）。

```
[root@ansible ansible]# vi lamp/roles/prepare/tasks/main.yml
- name: delete yum config
  shell: rm -rf /etc/yum.repos.d/*
- name: provide yumrepo file
  shell: wget -O /etc/yum.repos.d/CentOS-Base.repo http://mirrors.aliyun.com/repo/Centos-7.repo
- name: clean the yum repo
  shell: yum clean all
- name: clean the iptables
  shell: iptables -F
```

（4）构建 httpd 服务任务，编写 palybook 用于安装 httpd 服务和 PHP 服务测试页，详细信息如下所示：

```
[root@ansible ansible]# cd /etc/ansible/lamp/roles/
[root@ansible roles]# mv /var/www/html/index.php  httpd/files/
[root@ansible roles]# vi httpd/tasks/main.yml
- name: web server install
  yum: name=httpd state=present                   # 安装 httpd 服务
- name: provide test page
  copy: src=index.php dest=/var/www/html
- name: delete apache config
  shell: rm -rf  /etc/httpd/conf/httpd.conf
- name: provide configuration file
  copy: src=httpd.conf dest=/etc/httpd/conf/httpd.conf   # 提供测试页
  notify: restart httpd
```

扩展：notify 和 handlers

① notify：这个 action 可用于在每个 play 的最后被触发，这样可以避免多次有改变发生时，每次都执行指定的操作，取而代之，仅在所有的变化发生完成后一次性地执行指定操作。在 notify 中列出的操作称为 handler，也即在 notify 中调用 handler 中定义的操作。

② handlers：handlers 也是一些 task 的列表，通过名字来引用，它们和一般的 task 并没有什么区别。Handlers 是由通知者进行 notify，如果没有被 notify，handlers 不会执行。不管有多少个通知者进行了 notify，等到 play 中的所有 task 执行完成之后，handlers 也只会被执行一次。handlers 最佳的应用场景是用来重启服务，或者触发系统重启操作，除此以外很少用到。

下面是构建 httpd 中的 handlers，详细信息如下所示：

```
[root@ansible roles]# vi httpd/handlers/main.yml
- name: restart httpd
  service: name=httpd enabled=yes state=restarted
```

（5）构建 MariaDB 数据库任务，首先创建 MySQL 服务的任务，需要安装 MySQL 服务，改变属主信息，然后启动 MySQL，详细信息如下所示：

```
[root@ansible roles]# vi mysql/tasks/main.yml
- name: install the mysql
  yum: name=mariadb-server state=present          # 安装 MySQL 服务
- name: mkdir date directory
  shell: mkdir -p /mydata/data                    # 创建挂载点目录
- name: provide configuration file
  copy: src=my.cnf dest=/etc/my.cnf               # 提供 MySQL 的配置文件
- name: change the owner
  shell: chown -R mysql:mysql /mydata/*           # 更改属主和属组
- name: start mariadb
  service: name=mariadb enabled=yes state=started # 启动 MySQL 服务
```

（6）构建 PHP 的任务，首先要安装 PHP 服务，然后安装数据库插件，详细信息如下所示：

```
[root@ansible roles]# vi php/tasks/main.yml
- name: install php
  yum: name=php state=present                     # 安装 PHP
- name: install php-mysql
  yum: name=php-mysql state=present               # 安装 PHP 与 MySQL 交互的插件
```

（7）配置 site.yml 文件作为执行入口文件，文件中一般用于定义所使用的 roles 操作，详细信息如下所示：

```
[root@ansible roles]# vi site.yml
- name: LAMP build
  remote_user: root
  hosts: web-servers
  roles:
    - prepare
    - mysql
    - php
    - httpd
```

3. 服务部署与测试

（1）使用 ansible-playbook 命令部署服务，命令如下所示，局部结果如图 7-29 所示。

```
[root@ansible roles]# ansible-playbook -i /etc/ansible/hosts
/etc/ansible/lamp/roles/site.yml
```

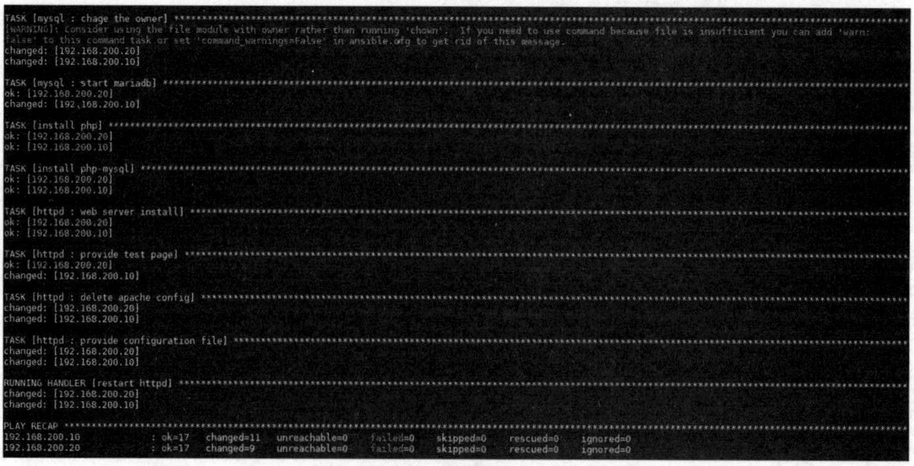

图 7-29　服务部署成功

第 7 章 监控服务与自动化运维工具

（2）服务编排完成之后，在浏览器中输入 ansible 和 cient 节点的 IP，即 192.168.200.10 和 192.168.200.20，结果如图 7-30、图 7-31 所示，至此完成了 Playbook 批量部署 LAMP 环境。

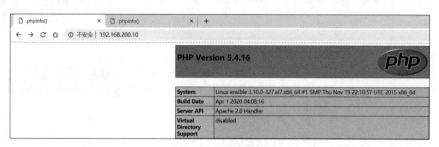

图 7-30　网页测试

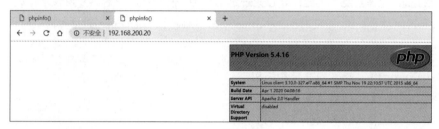

图 7-31　网页测试

本任务到此结束。

小　　结

在本章中，您已经学会：
- Nagios 监控服务安装与部署。
- Zabbix 监控服务安装与部署。
- Ansible 自动化服务安装与部署。
- Ansible 批量部署 LAMP 环境。

第 8 章

容器高级运维

📝 本章概要

学习在 Linux 操作系统中通过部署 Docker 容器服务,使用 Docker 容器应用技术完成仓库配置、镜像构建、容器编排等知识,并使用 docker-compose 容器编排技术完成 gpmall 应用商城的容器化部署。

学习目标

- 学习 Docker 容器技术的原理和架构。
- 掌握 Docker 容器的安装与仓库配置。
- 掌握 Docker 容器的基础管理。
- 掌握 Dockerfile 编写。
- 掌握 Docker 容器编排。
- 掌握容器应用商城系统编排。

思维导图

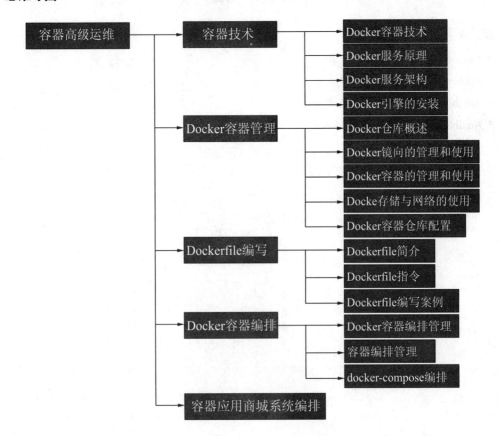

🛡 任务目标

学习 Docker 容器应用技术的部署与管理，通过学习相应模块，完成容器应用商城系统的编排。

💡 8.1 容器技术

学习目标

学完本节后，您应能够：
- 了解 Docker 服务背景。
- 了解 Docker 服务原理。
- 了解 Docker 服务架构。

8.1.1 Docker 容器技术

1. Docker 容器的发展

当今，云计算技术正在成为信息技术产业发展的战略重点，全球的信息技术企业都在纷纷向云计算转型。而云计算的服务模式仍在不断进化，但业界普遍认为云计算按照服务的提供方式划分为基础设施计算资源（IaaS）、平台（PaaS）、软件（SaaS）。

Docker 自开源后受到广泛的关注和讨论，至今其 GitHub 项目已经超过 36 000 千个 Stat（星标）和一万多个 Fork（分支）。甚至由于 Docker 项目的火爆，在 2013 年底，DotCloud 公司决定改名为 Docker。Docker 最初是在 Ubuntu 12.04 上开发实现的；Red Hat 则从 RHEL6.5 开始对 Docker 进行支持；Google 也在其 PaaS 产品中广泛应用 Docker。

2. Docker 容器简介

容器是一种轻量级的、可移植的、自包含的软件打包技术，使应用程序几乎可以在任何地方以相同的方式运行。开发人员在自己的笔记本计算机上创建并测试好的容器，无须任何修改就能够在生产系统的虚拟机、物理服务器或公有云主机上运行。容器的本质就是一种基于操作系统能力的隔离技术，是一组受到资源限制且彼此间相互隔离的进程。运行这些进程所需要的所有文件都由另一个镜像提供，也就意味着从开发到测试再到生产的整个过程中，容器都具有可移植性和一致性。容器自身没有操作系统，而是直接共享宿主机的内核，所有对于容器进程的限制都是基于操作系统本身的能力来进行的。因此，容器最大的优势就是轻量化。

谈到容器就不得不提其与虚拟机技术的区别。传统虚拟机技术是虚拟了一套硬件后，在其上运行一个完整的操作系统，在该系统上再运行所需应用进程。而容器则可共享同一个操作系统的内核，将应用进程与系统其他部分隔离开。

如图 8-1 所示可以看出，容器与虚拟机之间的主要区别在于虚拟化层的位置和操作系统资源的使用方式。虚拟化会使用虚拟机监控程序模拟硬件，从而使多个操作系统能够并行运行；Linux 容器则是在本机操作系统上运行，与所有容器共享该操作系统。因此，在资源有限的情况下，想要进行密集部署的轻量级应用时，容器技术就能凸显出其优势。与虚拟机相比，更重要的是 Linux 容器在运行时所占用的资源更少，使用的是标准接口（启动、停止、环境变量等），并且会与应用相隔离。此外，作为包含多个容器大型应用的一部分时，更加易于管理，而且这些多容器应用可以跨多个云环境进行编排。

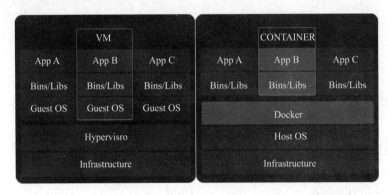

图 8-1　容器与虚拟化

8.1.2　Docker 服务原理

1. Docker 容器的原理

Docker 使用 Google 公司推出的 Go 语言进行开发实现，基于 Linux 内核的 Cgroups、Namespace 以及 AUFS 类的 UnionFS 等技术，对进程进行封装隔离，属于操作系统层面的虚拟化技术。最初实现是基于 LXC，从 Docker 0.7 以后开始去除 LXC，转而使用自行开发的 Libcontainer，从 Docker 1.11 开始，则进一步演进为使用 RunC 和 Containerd。

（1）Cgroups 即为 Control groups，其作用就是在 Linux 中限制某个或者某些进程的分配资源。在 group 中，有分配好的特定比例的 CPU 时间、IO 时间、可用内存大小等。Cgroups 是将任意进程进行分组化管理的 Linux 内核功能。最初由 Google 的工程师提出，后来被整合进 Linux 内核中。Cgroups 中的重要概念是"子系统"，也就是资源控制器，每个子系统就是一个资源的分配器。Cgroups 被 Linux 内核支持，有得天独厚的性能优势，发展势头迅猛。在很多领域可以取代虚拟化技术分割资源。Cgroup 默认有诸多资源组，可以限制几乎所有服务器上的资源，如 cpu、mem iops、iobandwide、net、device access 等。

（2）Namespace。（命名空间）是 Linux 中用于分离进程树、网络接口、挂载点以及进程间通信等资源的方法。Linux 主要有 7 种不同的命名空间，包括 CLONE_NEWCGROUP、CLONE_NEWIPC、CLONE_NEWNET、CLONE_NEWNS、CLONE_NEWPID、CLONE_NEWUSER 和 CLONE_NEWUTS，通过这 7 个选项能在创建新的进程时设置新进程应该在哪些资源上与宿主机器进行隔离。Docker 就是通过 Linux 的 Namespaces 对不同的容器实现隔离的。

（3）AUFS（Another Union File System）是一个能透明覆盖一或多个现有文件系统的层状文件系统，支持将不同目录挂载到同一个虚拟文件系统下的文件系统，可以把不同的目录联合在一起，组成一个单一的目录。这是一种虚拟的文件系统，文件系统不用格式化，直接挂载即可。Docker 则一直在用 AUFS 作为容器的文件系统。当一个进程需要修改一个文件时，AuFS 创建该文件的一个副本。AUFS 可以把多层合并文件系统的单层表示。这个过程称为写入复制（copy on write）。AUFS 允许 Docker 把某种镜像作为容器的基础。使用 AuFS 的另一个好处是 Docker 的版本容器镜像能力，每个新版本都是一个与之前版本的简单差异改动，有效地保持镜像文件最小化。

因此，基于 Linux 命名空间（Namespaces）、控制组（Cgroups）和 AUFS 三大技术才支撑了目前 Docker 的实现，也是 Docker 能够出现的最重要的原因。

2. Docker 容器的优势

相较传统的虚拟化方式，Docker 主要有以下几方面的优势：

（1）更高效地利用系统资源。Docker 对系统资源的利用率很高，一台主机上可以同时运行数千个 Docker 容器。容器除了运行其中应用外，基本不消耗额外的系统资源，使得应用的性能很高，同时系统的开销尽量小。

（2）更快速地交付和部署。Docker 在整个开发周期中都可以辅助实现快速交付，并且允许开发者在装有应用和服务的本地容器做开发，可以直接集成到可持续开发流程之中。

（3）更高效地部署和扩容。Docker 容器几乎可以运行于任意平台上，包括物理机、虚拟机、公有云、私有云等，这种兼容性就非常方便用户把一个应用程序从一个平台直接迁移到另外一个平台。Docker 的兼容性和轻量特性可以很轻松地实现负载动态管理，可以快速扩容或方便下线应用和服务。

（4）更简单的管理。使用 Docker，通常只需要小小的改变就可以替代以往大量的更新工作。所有的修改都是以增量的方式被分发和更新，从而实现自动化且高效的管理。

8.1.3　Docker 服务架构

1. Docker 容器架构

Docker 使用客户端/服务器（C/S）架构模式，使用远程 API 来管理和创建 Docker 容器。Docker 客户端只需向 Docker 服务器或守护进程发出请求，服务器或守护进程将完成所有工作并返回结果。Docker 提供了一个命令行工具 docker 以及一整套 RESTful API 进行通信，可以在同一台宿主机上运行 Docker 守护进程和客户端，也可以从本地的 Docker 客户端连接到运行在另一台宿主机上的远程 Docker 守护进程。如图 8-2 所示为 Docker 服务的架构图。

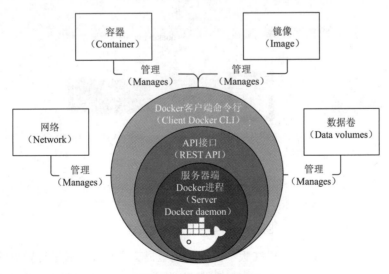

图 8-2　Docker 架构图

2. Docker 的组件

一个完整的 Docker 服务包括 Docker Daemon 服务器、Docker Client 客户端、Docker Image

镜像、Docker Registry 库和 Docker Contrainer 容器，如图 8-3 所示。

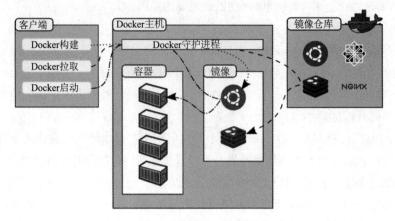

图 8-3　Docker 的组件

（1）Docker 镜像。

Docker 镜像是一个只读模板，用于创建 Docker 容器，由 Dockerfile 文本描述镜像的内容。构建一个镜像实际就是安装、配置和运行的过程。Docker 镜像基于 UnionFS 把以上过程进行分层（Layer）存储，这样更新镜像可以只更新变化的层。

Docker 镜像有多种生成方法：

① 可以从无到有开始创建镜像。

② 可以下载并使用别人创建好的现成的镜像。

③ 可以在现有镜像上创建新的镜像。

Docker Hub 提供了很多镜像，但在实际工作中，Docker Hub 中的镜像并不能满足工作的需要，往往需要构建自定义镜像。构建自定义镜像主要有两种方式：docker commit 和 Dockerfile，如图 8-4 所示。

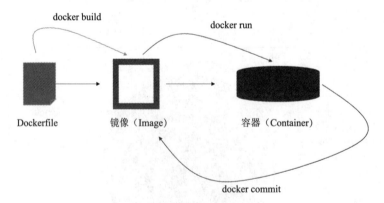

图 8-4　构建自定义镜像方式

可以将 docker commit 视为在以往版本控制系统里提交变更，然后进行变更的提交即可。docker commit、docker export 和 docker add 等都可以输出 image，但是最好的生成 image 的方法还是使用 Dockerfile。

Dockerfile 是由一系列命令和参数构成的脚本，这些命令应用于基础镜像并最终创建一个新的镜像。它们简化了从头到尾的流程并极大地简化了部署工作。Dockerfile 从 FROM 命令开始，紧接着跟随各种方法、命令和参数，其产出为一个新的可以用于创建容器的镜像。

（2）Docker 容器。

Docker 容器是一个镜像的运行实例。它可以被启动、开始、停止和删除。每个容器都是相互隔离的、保证安全的平台。Dcoker 容器由应用程序本身和依赖两部分组成。容器在宿主机操作系统的用户空间中运行，与操作系统的其他进程隔离。这一点显著区别于的虚拟机。

（3）Docker 仓库。

Docker 仓库是 Docker 镜像库，是用来集中存放镜像文件的场所。Docker Registry 也是一个容器，往往存放着多个仓库，每个仓库中又包含了多个镜像，每个镜像有不同的标签（tag）。Docker Hub 是 Docker 公司提供的互联网公共镜像仓库，用户可以在上面找到需要的镜像，也可以把私有镜像推送上去。但是，在生产环境中往往需要一个私有的镜像仓库用于管理镜像，通过开源软件 Registry 可以实现这个目标。

8.1.4　任务：Docker 引擎的安装

Docker 使用客户端/服务器（C/S）架构模式，使用远程 API 来管理和创建 Docker 容器。Docker 客户端只需向 Docker 服务器或守护进程发出请求，服务器或守护进程将完成所有工作并返回结果。

任务执行清单

在本任务中，您将通过实践部署 Docker 引擎和系统。

目标

- 了解 Docker 引擎和系统架构。
- 掌握 Docker 引擎的部署和配置。

重要信息

- 本任务采用 CentOS 7.5_1804 操作系统。
- 虚拟机密码为 000000。
- 关闭防火墙和 selinux。

解决方案

1. 基础环境配置

（1）在 Vmware 中安装两台虚拟机，分别为 master 和 slave 节点，下面先登录两个节点更改主机名和主机映射关系。

```
[root@localhost ~]# hostnamectl set-hostname master
[root@localhost ~]# logout
[root@master ~]#
[root@localhost ~]# hostnamectl set-hostname slave
[root@localhost ~]# logout
[root@slave ~]#
[root@master ~]# vi /etc/hosts
127.0.0.1    localhost localhost.localdomain localhost4 localhost4.localdomain4
::1          localhost localhost.localdomain localhost6 localhost6.localdomain6
192.168.100.10 master
192.168.100.20 slave
```

```
[root@slave ~]# vi /etc/hosts
127.0.0.1     localhost localhost.localdomain localhost4 localhost4.localdomain4
::1           localhost localhost.localdomain localhost6 localhost6.localdomain6
192.168.100.10 master
192.168.100.20 slave
```

（2）将 PaaS.iso 镜像包上传至 master 节点，在 master 节点创建两个文件夹，并将镜像挂载到该目录下，命令和结果如下所示：

```
[root@master ~]# mkdir /opt/{centos,docker}
[root@master ~]# mount -o loop paas.iso /opt/docker/
mount: /dev/loop0 is write-protected, mounting read-only
[root@master ~]# mount /dev/cdrom /opt/centos/
mount: /dev/sr0 is write-protected, mounting read-only
[root@master ~]# df -h
Filesystem               Size  Used Avail Use% Mounted on
/dev/mapper/centos-root   36G  9.5G   27G  27% /
devtmpfs                 903M     0  903M   0% /dev
tmpfs                    913M     0  913M   0% /dev/shm
tmpfs                    913M  8.6M  904M   1% /run
tmpfs                    913M     0  913M   0% /sys/fs/cgroup
/dev/sda1                497M  114M  384M  23% /boot
tmpfs                    183M     0  183M   0% /run/user/0
/dev/loop0               8.7G  8.7G     0 100% /opt/docker
/dev/sr0                 4.2G  4.2G     0 100% /opt/centos
```

（3）在 master 节点配置本地 yum 源文件，命令和方法如下所示：

```
[root@master ~]# mv /etc/yum.repos.d/* /home/
[root@master ~]# vi /etc/yum.repos.d/local.repo
[docker]
name=docker
baseurl=file:///opt/docker/kubernetes-repo
enabled=1
gpgcheck=0
[centos]
name=centos
baseurl=file:///opt/centos
enabled=1
gpgcheck=0
```

（4）安装 FTP 服务，使得 slave 节点可以直接使用文件共享，随即在 slave 节点配置 yum 文件。

```
[root@master ~]# yum -y install vsftpd
[root@master ~]# vi /etc/vsftpd/vsftpd.conf
anon_root=/opt
[root@master ~]# systemctl restart vsftpd
```

（5）在 master 节点和 slave 节点开启路由转发功能，此处以 master 节点为例，命令和结果如下所示：

```
[root@master ~]# vi /etc/sysctl.conf
net.ipv4.ip_forward=1
net.bridge.bridge-nf-call-ip6tables = 1
net.bridge.bridge-nf-call-iptables = 1
[root@master ~]# modprobe br_netfilter
[root@master ~]# sysctl -p
```

```
net.ipv4.ip_forward = 1
net.bridge.bridge-nf-call-ip6tables = 1
net.bridge.bridge-nf-call-iptables = 1
```

2. 引擎安装

(1) yum-utils 提供了 yum-config-manager 的依赖包,device-mapper-persistent-data 和 lvm2 需要 devicemapper 存储驱动,slave 节点进行相同步骤。

```
[root@master ~]# yum install -y yum-utils device-mapper-persistent-data lvm2
```

(2) 随着 Docker 的不断流行与发展,Docker 组织也开启了商业化之路,Docker 从 17.03 版本之后分为 CE(Community Edition)和 EE(Enterprise Edition)两个版本。

Docker EE 专为企业的发展和 IT 团队建立,为企业提供最安全的容器平台和以应用为中心的平台,有专门的团队支持,可在经过认证的操作系统和云提供商中使用,并可运行来自 DockerStore 的经过认证的容器和插件。

Docker CE 是免费的 Docker 产品的新名称,Docker CE 包含了完整的 Docker 平台,非常适合开发人员和运维团队构建容器 App。

此处安装 Docker CE 版本,然后启动并设置开机自启,命令如下所示,slave 节点进行相同步骤。

```
[root@master ~]# yum install -y docker-ce containerd.io
[root@master ~]# systemctl daemon-reload
[root@master ~]# systemctl restart docker
[root@master ~]# systemctl enable docker
Created symlink from /etc/systemd/system/multi-user.target.wants/docker.service to /usr/lib/systemd/system/docker.service.
```

(3) 在 master 节点查看 Docker 的详细系统信息,结果如下所示:

```
[root@master ~]#  docker info
Client:
 Debug Mode: false
Server:
 Containers: 0
  Running: 0
  Paused: 0
  Stopped: 0
 Images: 0
 Server Version: 19.03.13
 Storage Driver: overlay2
  Backing Filesystem: xfs
  Supports d_type: true
  Native Overlay Diff: true
 Logging Driver: json-file
 Cgroup Driver: cgroupfs
```

本任务到此结束。

8.2 Docker 容器管理

学习目标

学完本节后，您应能够：
- 了解 Docker 容器的仓库。
- 了解 Docker 容器的管理。
- 掌握 Dockers 容器的仓库配置。

8.2.1 Docker 仓库概述

1. Docker 仓库

镜像构建完成后，可以很容易地在当前宿主机上运行。但是，如果需要在其他服务器上使用这个镜像，就需要一个集中的存储、分发镜像的服务器，Docker Registry（镜像注册）就是这样的服务。一个 Docker Registry 中可以包含多个仓库（Registry）；每个仓库可以包含多个标签（Tag）；每个标签对应一个镜像。

通常，一个仓库会包含一个软件不同版本的镜像，标签常用于对应该软件的各个版本。开发者可以通过"< 仓库名 >:< 标签 >"的格式来指定具体是这个软件哪个版本的镜像。如果不给出标签，则以 Latest 作为默认标签。

仓库名经常以两段式路径形式出现，如 james/php_web，前者往往是 Docker Registry 多用户环境下的用户名，后者则往往是对应的软件名。但这并非绝对，主要取决于所使用的具体 Docker Registry 的软件或服务。

Docker Registry 服务可以分为两种。一种为公开并开放给所有的用户使用，包含用户的搜索、拉取，镜像提交时更新，还可以免费保管用户镜像数据。这类服务受制于网络带宽的限制，并不能及时、快速地获取所需要的资源，但是优点是可以获取大部分并可以立即使用的镜像，减少镜像的制作时间。另外一种服务是在一定范围对特定的用户提供 Registry 服务，一般存在于学校内部、企业内部的服务管理和研发等环境，这在一定程度上保证了镜像拉取的速度，对内部核心镜像数据有保护作用，但是也存在镜像内容不丰富的问题。

2. 私有仓库

（1）私有仓库的特点

仓库（Registry）是集中存放镜像的地方，在上节内容已经说明了 Docker 仓库分为公有仓库和私有仓库，然而公有仓库在某些情况下并不适用于公司内部传输。通过对比两种仓库的特点，大致可以得出私有仓库具有节省带宽、传输速度快、方便存储的优点。

（2）Docker Registry 工作方式

Docker Registry 是 Image 的仓库，当开发者编译完成一个 Image 时，就可以推送到公共的 Registry，如 Docker Hub，也可以推送到自己的私有 Registry。使用 Docker Client，开发者可以搜索已经发布的 Image，从中拉取 Image 到本地，并在容器中运行。

Docker Hub 提供了公有和私有的 Registry。所有人都可以搜索和下载公共镜像，私有仓库只有私有用户才能查询和下载。

8.2.2 Docker 镜像的管理和使用

1. Docker 镜像

Docker 所宣称的用户可以随心所欲地"Build、Ship and Run"应用的能力，其核心是由 Docker image（Docker 镜像）来支撑的。Docker 通过把应用的运行时环境和应用打包在一起，解决了部署环境依赖的问题；通过引入分层文件系统这种概念，解决了空间利用的问题。它彻底消除了编译、打包与部署、运维之间的鸿沟，与现在互联网企业推崇的 DevOps 理念不谋而合，大大提高了应用开发部署的效率。Docker 公司的理念被越来越多的人理解和认可也就是理所当然的了，而理解 Docker image 则是深入理解 Docker 技术的一个关键点。

2. Docker image

简单地说，Docker image 是用来启动容器的只读模板，是容器启动所需要的 rootfs，类似于虚拟机所使用的镜像。首先需要通过一定的规则和方法表示 Docker image，如图 8-5 所示。

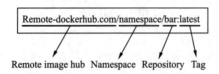

图 8-5　Docker 镜像的典型表示法

图 8-5 是典型的 Docker 镜像的表示方法，可以看到其被"/"分为了三个部分，其中每部分都可以类比 Github 中的概念。下面按照从左到右的顺序介绍这几个部分以及一些相关的重要概念。

（1）Remote docker hub：集中存储镜像的 Web 服务器地址。该部分的存在使得可以区分从不同镜像库中拉取的镜像。若 Docker 的镜像表示中缺少该部分，说明使用的是默认镜像库，即 Docker 官方镜像库。

（2）Namespace：类似于 Github 中的命名空间，是一个用户或组织中所有镜像的集合。

（3）Repository：类似于 Git 仓库，一个仓库可以有多个镜像，不同镜像通过 tag 来区分。

（4）Tag：类似 Git 仓库中的 tag，一般用来区分同一类镜像的不同版本。

（5）Layer：镜像由一系列层组成，每层都用 64 位的十六进制数表示，非常类似于 Git 仓库中的 commit。

（6）Image ID：镜像最上层的 layer ID 就是该镜像的 ID，Repo:tag 提供了易于人类识别的名字，而 ID 便于脚本处理、操作镜像。

镜像库是 Docker 公司最先提出的概念，非常类似应用市场的概念。用户可以发布自己的镜像，也可以使用别人的镜像。Docker 开源了镜像存储部分的源代码（Docker Registry 以及 Distribution），但是这些开源组件并不适合独立地发挥功能，需要使用 Nginx 等代理工具添加基本的鉴权功能，才能搭建出私有镜像仓库。本地镜像则是已经下载到本地的镜像，可以使用 docker images 等命令进行管理。这些镜像默认存储在 /var/lib/docker 路径下，该路径也可以使用 docker daemon –g 参数在启动 Daemon 时指定。

> **注意**：Docker 的镜像已经支持更多层级，比如用户的命名空间之前可以包含组织（Remote-dockerhub.com/group/namespace/bar:latest）。但是目前 Docker 官方的镜像库还不具备该能力。

3. 镜像的管理

（1）docker load 命令，可导入镜像文件 *.tar，该文件一般是由 docker save 命令进行镜像导

出的,其中 -i 和 < 表示从文件输入。会成功导入镜像及相关元数据,包括 tag 信息。下面为操作实例:

```
[root@master ~]# docker load -i /opt/docker/images/centos-centos7.5.1804.tar
4826cdadf1ef: Loading layer [================================================
=====>]  207.8MB/207.8MB
Loaded image: centos:centos7.5.1804
```

(2)docker images 命令,下面的命令可以列出本地存储中的镜像列表,也可以查看这些进行的基本信息。

```
[root@master ~]# docker images
REPOSITORY          TAG                 IMAGE ID            CREATED             SIZE
centos              centos7.5.1804      cf49811e3cdb        2 years ago         200MB
```

① REPOSITORY:表示镜像的仓库源。
② TAG:镜像的标签,用于区分同一个仓库中的不同镜像。
③ IMAGE ID:镜像的唯一标识:64 位 HashID。
④ CREATED:镜像创建时间。
⑤ SIZE:镜像所占用的虚拟大小,该大小包含了所有共享文件的大小。

同一仓库源可以有多个 TAG,代表这个仓库源的不同版本。例如,http 仓库源里有 2.2.31、2.2.32 等多个不同的版本,可以使用 REPOSITORY:TAG 命令来定义不同的镜像。

(3)docker push 命令,首先使用 docker tag 将镜像打上标签,然后使用 docker push 命令上传标记的镜像到仓库。

```
[root@master ~]# docker tag cf49811e3cdb 192.168.100.10:5000/centos:1804
[root@master ~]# docker push 192.168.100.10:5000/centos:1804
The push refers to repository [192.168.100.10:5000/centos]
4826cdadf1ef: Pushed
1804: digest: sha256:65decb5f8c6d37cdd06332ef1116a92fdb52aa1b55fe6256bb3b8
43ee97d2279 size: 529
```

(4)docker pull 命令,当本地主机上使用一个不存在的镜像时,Docker 会自动下载这个镜像。如果需要预先下载这个镜像,可以使用 docker pull 命令来下载。

```
[root@slave ~]# docker pull 192.168.100.10:5000/centos:1804
[root@slave ~]# docker images
REPOSITORY                      TAG         IMAGE ID            CREATED             SIZE
192.168.100.10:5000/centos      1804        cf49811e3cdb        2 years ago         200MB
```

(5)删除镜像

删除镜像使用 docker rmi 命令,语法如下:

```
# docker rmi [OPTIONS] IMAGE [IMAGE...]
```

OPTIONS 说明:

```
l -f: 强制删除。
l --no-prune: 不移除该镜像的过程镜像,默认移除。
```

例如,强制删除本地镜像 centos。

```
[root@master ~]# docker rmi -f centos:centos7.5.1804
Untagged: centos:centos7.5.1804
```

8.2.3 Docker 容器的管理和使用

1. Container 容器概念

容器是镜像的另一个运行实例,是独立运行的一个或一组应用以及它们所必需的运行环境,包括文件系统、系统库类、shell 环境等。镜像是只读模板,而容器会给这个只读模板一个额外的可写层。

容器是一个镜像的运行实例,容器由镜像创建,运行用户指定的指令或者 Dockerfile 定义的运行指令,可以将其启动、停止、删除,而这些容器都是相互隔离(独立进程)、互不可见的。

2. 容器的管理

(1) docker run 命令,运行第一个容器,执行以下命令:

```
[root@master ~]# docker run -itd --rm --name centos
192.168.100.10:5000/centos:1804 /bin/bash
```

如果不指定镜像的版本标签,则默认使用 latest 标签的镜像。

参数说明:

- -i:交互式操作。
- -t:终端。
- -d:容器在后台运行。
- --rm:容器退出后即将其删除,可以避免浪费空间。
- --name:定义 container 容器名称。
- 192.168.100.10:5000/centos:1804:镜像名,使用仓库中 centos 镜像为基础来启动容器。
- /bin/bash:载入容器后运行 bash。

(2) docker ps 命令,可以查看当前状态下已经启动成功的容器,命令和结果如下所示:

```
[root@master ~]# docker ps -a
  CONTAINER ID            IMAGE                       COMMAND
CREATED                   STATUS                  PORTS                     NAMES
  54882f260545            192.168.100.10:5000/centos:1804      "/bin/bash"
About a minute ago        Up About a minute                                 centos
```

(3) docker exec 命令,在使用 docker ps 看到容器创建成功之后可以使用该命令进入容器,此时的容器就是一个 CentOS 7.5 的系统,最后输入 exit 或者按【Ctrl+C】组合键即可退出容器,示例代码如下所示:

```
[root@master ~]# docker exec -it centos /bin/bash
[root@54882f260545 /]# cd root/
[root@54882f260545 ~]# mkdir file
[root@54882f260545 ~]# ls
anaconda-ks.cfg    file
[root@54882f260545 ~]# exit
exit
```

(4) docker 容器的停止、启动、终止、删除等。

① 停止容器的语法如下:

```
docker stop [OPTIONS] CONTAINER [CONTAINER...]
```

② 启动容器的语法如下:

```
# docker start [CONTAINER ID]
```

③ 例如，启动所有的 Docker 容器。

```
# docker start $(docker ps -aq)
```

④ 删除运行中的容器。

```
# docker rm -f [CONTAINER ID]
```

⑤ 批量删除所有的容器。

```
# docker rm $(docker ps -aq)
```

⑥ 终止容器进程，容器进入终止状态。

```
# docker container stop [CONTAINER ID]
```

⑦ 将容器快照导出为本地文件，语法如下：

```
# docker export [CONTAINER ID] > [tar file]
```

8.2.4 Docker 存储与网络的使用

1. Docker 卷管理基础

数据卷是一个可供一个或多个容器使用的特殊目录，它绕过 UFS，可以提供很多有用的特性：

- 数据卷可以在容器之间共享和重用。
- 对数据卷的修改会立即生效。
- 对数据卷的更新不会影响镜像。
- 数据卷默认会一直存在，即使容器被删除。
- 数据卷的使用，类似于 Linux 下对目录或文件进行 mount，镜像中的被指定为挂载点的目录中的文件会隐藏掉，能显示看到的是挂载的数据卷。

（1）增加新数据卷。

用户可以在执行 docker create 或者 docker run 命令时使用 -v 参数来添加数据卷，也可以通过多次指定该参数来挂载多个数据卷。这里以创建 centos 容器为例。

```
[root@master ~]# docker run -itd -v /tmp/data --name nginx nginx:latest
6cbdeadb27750f58c4f3a7358169d7daff6ff551505cba32212e124caecf8f62
```

其中，-v 参数会在容器的 /tmp/data 目录下创建一个新的数据卷。

用户可以通过 docker inspect 命令查看数据卷在主机中的位置：

```
[root@master ~]# docker inspect nginx
…
"Mounts": [
        {
            "Type": "volume",
            "Name": "ec48666aac21c187a7e9179d42551ea96767675cbe97ffd143ccdd18cad7eba4",
            "Source": "/var/lib/docker/volumes/ec48666aac21c187a7e9179d42551ea96767675cbe97ffd143ccdd18cad7eba4/_data",
            "Destination": "/tmp/data",
            "Driver": "local",
            "Mode": "",
            "RW": true,
            "Propagation": ""
        }
    ],
…
```

（2）将主机目录挂载为数据卷

-v 参数除了可以用于创建数据卷外，还可以用来将 Docker daemon 所在主机上的文件或文件夹挂载到容器中，-v 参数的主机目录必须是绝对路径，如果指定路径不存在，Docker 会自动创建该目录，还可以使用只读方式挂载一个数据卷，如下所示：

```
[root@master ~]# docker run -it -v /host/data:/tmp/data:ro nginx:latest
[root@b7d52c675f37 /]# echo 123 > /tmp/data/id
bash: /tmp/data/id: Read-only file system
```

（3）创建数据卷容器

数据卷容器（Data volume containers）涉及容器间共享的持久化、序列化的数据持久性的数据，最好创建数据卷容器。数据卷容器，其实就是一个正常的容器，专门用来提供数据卷供其他容器挂载。

例如想要创建一个 mysql 数据卷容器，并且希望这些数据库之间共享数据，可以先运行一个数据库容器，命令如下所示：

```
[root@slave ~]# docker run -itd -v /mysqldb --name mysqldb
192.168.100.10:5000/mysql:5.6 /bin/bash
```

使用 --volumes-from 参数将上面生成的数据卷挂载进来，之后启动容器，各个容器之间就可以通过 mysqldb 数据卷共享数据了，命令如下所示：

```
[root@slave ~]# docker run -d --volumes-from mysqldb --name nginxdb
192.168.100.10:5000/nginx:latest
e80dca25b81aa1bef19b05aa7e55c0e2b884bff2f7cfcbe3178421bdce72f9e6
```

进入 nginxdb 容器可以查看到 mysqldb 已经挂载成功，然后进入该数据卷创建一个 nginxtest 测试文件夹，在 mysqldb 容器中可以检测到已完成数据卷共享。

```
[root@slave ~]# docker exec -it nginxdb /bin/bash
root@e80dca25b81a:/# df -h
Filesystem               Size  Used Avail Use% Mounted on
overlay                  300G  2.2G  298G   1% /
tmpfs                     64M     0   64M   0% /dev
tmpfs                    992M     0  992M   0% /sys/fs/cgroup
shm                       64M     0   64M   0% /dev/shm
/dev/mapper/centos-root  300G  2.2G  298G   1% /mysqldb
tmpfs                    992M     0  992M   0% /proc/asound
tmpfs                    992M     0  992M   0% /proc/acpi
tmpfs                    992M     0  992M   0% /proc/scsi
tmpfs                    992M     0  992M   0% /sys/firmware
root@e80dca25b81a:/# cd mysqldb/
root@e80dca25b81a:/mysqldb# ls
root@e80dca25b81a:/mysqldb# mkdir nginxtest
root@e80dca25b81a:~# exit
exit
[root@slave ~]# docker exec -it mysqldb /bin/bash
root@43efdf141a9a:/# cd mysqldb/
root@43efdf141a9a:/mysqldb# ls
nginxtest
root@43efdf141a9a:/mysqldb#
root@43efdf141a9a:/mysqldb# exit
exit
```

（4）删除数据卷

数据卷是被设计用来持久化数据的，它的生命周期独立于容器，Docker不会在容器被删除后自动删除数据卷，并且也不存在垃圾回收这样的机制来处理没有任何容器引用的数据卷。如果需要在删除容器的同时移除数据卷。可以在删除容器的时候使用docker rm -v命令。

2. Docker网络管理基础

Docker的网络实现其实就是利用了Linux上的网络命名空间、网桥和虚拟网络设备（VETH）等实现。默认Docker安装完成后会创建一个网桥docker0。Docker中的网络接口默认都是虚拟的网络接口。Docker容器网络在本地主机和容器内分别创建一个虚拟接口，并让它们彼此连通。

- --net=bridge：是默认值，连接到默认的网桥。
- --net=host：告诉Docker不要将容器网络放到隔离的命名空间中，使用本地主机的网络，它拥有完全的本地主机接口访问权限。
- --net=container:NAME_or_ID：让Docker将新建容器的进程放到一个已存在容器的网络栈中，新容器进程有自己的文件系统、进程列表和资源限制，但会和已存在的容器共享IP地址和端口等网络资源，两者进程可以直接通过lo环回接口通信。
- --net=none：让Docker将新容器放到隔离的网络栈中，但是不进行网络配置。

（1）网络配置

用户使用--net=none后，可以自行配置网络，让容器达到跟平常一样具有访问网络的权限。通过这个过程，可以了解Docker配置网络的细节。

启动一个centos的容器，使用docker inspect命令查看容器详细信息，发现此时的容器是没有任何网络信息的。

```
[root@slave ~]#docker  run   -itd  --net=none   --name centos 192.168.100.10:5000/centos:1804
889a93cb3b48d46dd8a6a0603c22a651573253701c24bf6c41ad7f44691191ed
[root@slave ~]# docker inspect centos
"Networks": {
            "none": {
                "IPAMConfig": null,
                "Links": null,
                "Aliases": null,
                "NetworkID": "5552bf7a689337bb2b0fb9dce0f1f1839088c2dcfd509aa72fe411e668aaf886",
                "EndpointID": "323f73b7c97425a95ede29c579b07ca2018105c582a1c2d22a027e6a750f3cbf",
                "Gateway": "",
                "IPAddress": "",
                "IPPrefixLen": 0,
                "IPv6Gateway": "",
                "GlobalIPv6Address": "",
                "GlobalIPv6PrefixLen": 0,
                "MacAddress": "",
                "DriverOpts": null
            }
```

8.2.5 任务：Docker 容器仓库配置

目前世界上最大最知名的公共仓库之一是 Docker 官方发布的 Docker Hub，上面有超过 15 000 个镜像。大部分需求都可以通过 Docker Hub 直接下载镜像来实现。国内比较知名的 Docker 仓库社区有 Docker Pool、阿里云等。

任务执行清单

在本任务中，您将通过实践部署 Docker 容器私有仓库的配置。

目标

- 掌握 Docker 容器的基本管理。
- 掌握 Docker 容器仓库的配置。

重要信息

- 本任务采用 CentOS 7.5_1804 操作系统。
- 虚拟机密码为 000000。
- 提前配置好 yum 源仓库。
- 关闭防火墙和 selinux。
- 所有节点已安装好 docker-ce。

解决方案

1. Registry 仓库配置

（1）登录 master 节点，将提供的压缩包 Docker.tar.gz 上传至 /root 目录并解压。官方在 Docker Hub 上提供了 Registry 的镜像，可以直接使用该 Registry 镜像来构建一个容器，搭建私有仓库服务，操作如下所示：

```
[root@master ~]# tar -zxvf Docker.tar.gz
[root@master ~]# docker load -i images/registry_latest.tar
d9ff549177a9: Loading layer [=========================================
====>]  4.671MB/4.671MB
f641ef7a37ad: Loading layer [=========================================
=====>]  1.587MB/1.587MB
d5974ddb5a45: Loading layer [=========================================
=====>]  20.08MB/20.08MB
5bbc5831d696: Loading layer [=========================================
=====>]  3.584kB/3.584kB
73d61bf022fd: Loading layer [=========================================
=====>]  2.048kB/2.048kB
Loaded image: registry:latest
```

（2）将 Registry 镜像运行并生成一个容器。

```
[root@master ~]# docker run -d -v /opt/registry:/var/lib/registry -p 5000:5000
--restart=always --name registry registry:latest
abac6122833e4661a815ea4d085251328640bf06d2c93a5bb0cff65d2f4b7e53
```

Registry 服务默认会将上传的镜像保存在容器的 /var/lib/registry 中，将主机的 /opt/registry 目录挂载到该目录，即可实现将镜像保存到主机的 /opt/registry 目录。

Registry 容器启动后，打开浏览器输入地址 http://IP:5000/v2/，如果出现图 8-6 所示的情况说明 Registry 运行正常。

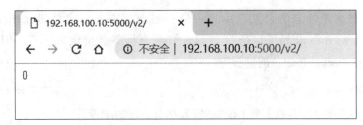

图 8-6　Register 仓库

（3）创建好私有仓库之后，就可以使用 docker tag 命令来标记一个镜像，然后推送它到仓库。先配置私有仓库，示例代码如下所示：

```
[root@master ~]# vi /etc/docker/daemon.json
{
  "insecure-registries": ["192.168.100.10:5000"]
}
[root@master ~]# systemctl restart docker
```

将 centos 镜像导入到本地存储中，然后使用 docker tag 命令将 centos 这个镜像标记为 192.168.100.10:5000/centos:1804。

```
[root@master ~]# docker load -i /opt/docker/images/centos-centos7.5.1804.tar
4826cdadf1ef: Loading layer [==================================================
=====>]  207.8MB/207.8MB
Loaded image: centos:centos7.5.1804
[root@master ~]# docker images
REPOSITORY     TAG              IMAGE ID         CREATED           SIZE
centos         centos7.5.1804   cf49811e3cdb     2 years ago       200MB
[root@master ~]# docker tag cf49811e3cdb 192.168.100.10:5000/centos:1804
```

使用 docker push 上传标记的镜像：

```
[root@master ~]# docker push 192.168.100.10:5000/centos:1804
The push refers to repository [192.168.100.10:5000/centos]
4826cdadf1ef: Pushed
1804: digest: sha256:65decb5f8c6d37cdd06332ef1116a92fdb52aa1b55fe6256bb3b843ee97d2279 size: 529
```

使用 curl 命令查看仓库中的镜像：

```
[root@master ~]# curl -L http://192.168.100.10:5000/v2/_catalog
{"repositories":["centos"]}
```

如同上述代码所示，提示 {"repositories":["centos"]}，则表明镜像已经上传成功，也可以在浏览器中访问上述地址查看，如图 8-7 所示。

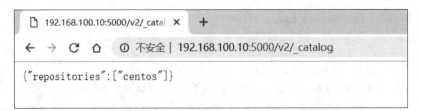

图 8-7　Register 仓库镜像

2. 拉取镜像

（1）登录 Slave 节点，配置私有仓库地址。

```
[root@slave ~]# vi /etc/docker/daemon.json
{
  "insecure-registries": ["192.168.100.10:5000"]
}
[root@slave ~]# systemctl restart docker
[root@slave ~]# docker info
……
 Insecure Registries:                                  #可以看见显示了仓库地址
 192.168.100.10:5000
 127.0.0.0/8
 Live Restore Enabled: false
```

（2）拉取镜像并查看结果，命令和结果如下所示：

```
[root@slave ~]# docker pull 192.168.100.10:5000/centos:1804
[root@slave ~]# docker images
192.168.100.10:5000/centos    1804    cf49811e3cdb    2 years ago    200MB
```

本任务到此结束。

8.3 Dockerfile 编写

学习目标

学完本节后，您应能够：
- 了解 Dockerfile 基础知识。
- 了解 Dockerfile 模块参数。
- 掌握 Dockerfile 案例编写。

8.3.1 Dockerfile 简介

Dockerfile 是一个包含用于组合映像的命令的文本文档。可以使用在命令行中调用任何命令。Docker 通过读取 Dockerfile 中的指令自动生成映像。

docker build 命令用于从 Dockerfile 构建映像。可以在 docker build 命令中使用 -f 标志指向文件系统中任何位置的 Dockerfile。

（1）Dockerfile 的基本结构

Dockerfile 一般分为四部分：基础镜像信息、维护者信息、镜像操作指令和容器启动时执行指令，"#"为 Dockerfile 中的注释。

（2）Dockerfile 文件说明

Docker 以从上到下的顺序运行 Dockerfile 的指令。为了指定基本映像，第一条指令必须是 FROM。一个声明以#字符开头则被视为注释。可以在 Docker 文件中使用 RUN、CMD、FROM、EXPOSE、ENV 等指令。

8.3.2 Dockerfile 指令

Dockerfile 是由一系列命令和参数构成的脚本，这些命令应用于基础镜像并最终创建一个新的镜像。它们简化了从头到尾的流程，并极大地简化了部署工作。Dockerfile 从 FROM 命令开始，紧接着跟随着各种方法、命令和参数、其产出为一个新的可以用于创建容器的镜像。

1. ROM 指令

FROM 是指定基础镜像，必须为第一个命令，格式为：

```
FROM <image>:<tag>
```

其中 tag 或 digest 是可选的，如果不使用这两个值时，会使用 latest 版本的基础镜像。

2. MAINTAINER 指令

MAINTAINER 用来声明维护者信息，格式为：

```
MAINTAINER <name>
```

3. LABEL 指令

LABEL：用于为镜像添加元数据，多用于声明构建信息，作者、机构、组织等。格式为：

```
LABEL <key>=<value> <key>=<value> <key>=<value> ...
```

使用 LABEL 指定元数据时，一条 LABEL 指令可以指定一或多条元数据，指定多条元数据时不同元数据之间通过空格分隔。推荐将所有的元数据通过一条 LABEL 指令指定，以免生成过多的中间镜像。

4. ENV 指令

ENV 用来设置环境变量，格式为：

```
ENV <key> <value>
ENV <key>=<value>
```

5. ARG 指令

ARG 用于指定传递给构建运行时的变量，格式为：

```
ARG <name>[=<default value>]
```

6. WORKDIR 指令

WORKDIR 用来指定工作目录，类似于通常使用的 cd 命令，格式为：

```
WORKDIR <PATH>
```

通过 WORKDIR 设置工作目录，Dockerfile 中的其他指令 RUN、CMD、ENTRYPOINT、ADD、COPY 等命令都会在该目录下执行。在使用 docker run 运行容器时，可以通过 -w 参数覆盖构建时所设置的工作目录。

7. ADD 指令

ADD 用于将本地文件添加到镜像中，tar 类型文件会自动解压（网络压缩资源不会被解压），可以访问网络资源，类似 wget，格式为：

```
ADD <src>... <dest>
```

8. COPY 指令

COPY 的功能类似于 ADD，但是不会自动解压文件，也不能访问网络资源。

9. RUN 指令

RUN 用来执行构建镜像时执行的命令，有以下两种命令执行方式：

（1）shell 执行格式为：
```
RUN <command>
```
（2）exec 执行格式为：
```
RUN ["executable", "param1", "param2"]
```

10. CMD 指令

CMD 构建容器后执行的命令，也就是在容器启动时才执行的命令。格式为：
```
# 执行可执行文件，优先执行
CMD ["executable","param1","param2"]
# 设置了 ENTRYPOINT，则直接调用 ENTRYPOINT 添加参数  参见 CMD 讲解
CMD ["param1","param2"]
# 执行 shell 命令
CMD command param1 param2
```

11. ENTRYPOINT 指令

ENTRYPOINT 用来配置容器，使其可执行化。配合 CMD 可省去 application，只使用参数。格式为：
```
# 可执行文件，优先
ENTRYPOINT ["executable", "param1", "param2"]
# shell 内部命令
ENTRYPOINT command param1 param2
```

12. EXPOSE 指令

EXPOSE 指定与外界交互的端口，格式为：
```
EXPOSE [<port>...]
```
EXPOSE 并不会直接让容器的端口映射主机。宿主机访问容器端口时，需要在 docker run 运行容器时通过 -p 来发布这些端口，或通过 -P 参数来发布 EXPOSE 导出的所有端口。

13. VOLUME 指令

VOLUME 用于指定持久化目录，格式为：
```
VOLUME ["<src>",...]
```
和 EXPOSE 指令类似，VOLUME 并不会挂载的宿主机，需要在 docker run 运行容器时通过 -v 来映射到宿主机的目录中。参见另一个命令 docker volume create。

14. USER 指令

USER 指定运行容器时的用户名或 UID，后续的 RUN 也会使用指定用户。使用 USER 指定用户时，可以使用用户名、UID 或 GID，或是两者的组合。当服务不需要管理员权限时，可以通过该命令指定运行用户。并且可以在之前创建所需要的用户，格式为：
```
USER user
USER user:group
USER uid:group
USER uid
USER user:gid
USER uid:gid
```
使用 USER 指定用户后，Dockerfile 中其后的命令 RUN、CMD、ENTRYPOINT 都将使用该用户。可以在 docker run 运行容器时，通过 -u 参数来覆盖指定用户。

15. ONBUILD 指令

ONBUILD 作用是当所构建的镜像被用作其他镜像的基础镜像时，该镜像中 ONBUILD 中的命令就会触发，格式为：

```
ONBUILD [INSTRUCTION]
```

8.3.3 任务：Dockerfile 编写案例

任务执行清单

在本任务中，您将通过实践编写 MySQL 数据库的镜像文件。

目标

- 掌握 Dockerfile 的指令。
- 掌握 Dockerfile 编写 MySQL。

Dockerfile 是一个纯文本文件，用以定义容器内部的安装配置环境，以构建出定制的容器镜像。

重要信息

- 本任务采用 CentOS 7.5_1804 操作系统。
- 虚拟机密码为 000000。
- 关闭防火墙和 selinux。
- 所有节点已安装好 docker-ce。

解决方案

1. Dockerfile 编写

（1）登录 master 节点，创建一个专门用于存放 mysql 镜像编写的文件夹，进入文件夹首先配置 yum 源文件，如下所示：

```
[root@master ~]# mkdir mysql
[root@master ~]# cd mysql/
[root@master mysql]# vi gpmall.repo
[docker]
name=docker
baseurl=ftp://192.168.100.10/docker/kubernetes-repo
enabled=1
gpgcheck=0
[centos]
name=centos
baseurl=ftp://192.168.100.10/centos
enabled=1
gpgcheck=0
[gpmall]
name=gpmall
baseurl=ftp://192.168.100.10/docker/gpmall-repo
enabled=1
gpgcheck=0
```

（2）编写 shell 脚本文件，用于构建 mysql 数据库，配置文件中包含了数据库基础配置信息，数据库文件以及数据库的远程访问权限等，详细配置如下所示：

```
[root@master mysql]# cat db_init.sh
```

```bash
#!/bin/bash
cat >> /etc/my.cnf << EOF
[mysqld]
port=3306
init_connect='SET collation_connection = utf8_unicode_ci'
init_connect='SET NAMES utf8'
character-set-server=utf8
collation-server=utf8_unicode_ci
skip-character-set-client-handshake
EOF
mysql_install_db --user=mysql
sleep 3
mysqld_safe &
sleep 3
#mysqladmin -u "$MARIADB_USER" -P 8066 password "$MARIADB_PASS"
mysql -e "use mysql; grant all privileges on *.* to '$MARIADB_USER'@'%' identified by '$MARIADB_PASS' with grant option;"
h=$(hostname)
mysql -e "use mysql; update user set password=password('$MARIADB_PASS') where user='$MARIADB_USER' and host='$h';"
mysql -e "flush privileges;"
mysql -e "create database gpmall;use gpmall;source /root/gpmall.sql;"
mysql -e "grant all privileges on *.* to '$MARIADB_USER'@'%' identified by '$MARIADB_PASS' with grant option;"
mysql -e "grant all privileges on *.* to 'root'@'%' identified by '123456' with grant option;"
mysql -e "grant all privileges on *.* to 'root'@'localhost' identified by '123456' with grant option;flush privileges;"
```

（3）编写 run.sh 启动文件为 Dockerfile 镜像文件中的默认启动命令。

```
[root@master mysql]# cat run.sh
#!/bin/bash
mysqld_safe
```

（4）mysql 数据库的 Dockerfile 如下所示：

```
[root@master mysql]# vi Dockerfile
FROM 192.168.100.10:5000/centos:1804
MAINTAINER Baicai
RUN rm -vf /etc/yum.repos.d/*
ADD local.repo /etc/yum.repos.d
RUN yum -y install mariadb-server

#设置环境变量，便于管理
ENV MARIADB_USER root
ENV MARIADB_PASS 123456
#让容器支持中文
ENV LC_ALL en_US.UTF-8

COPY gpmall.sql /root/
#初始化数据库
ADD db_init.sh /root/db_init.sh
RUN chmod 775 /root/db_init.sh
RUN /root/db_init.sh
```

```
# 导出端口
EXPOSE 3306

# 添加启动文件
ADD run.sh /root/run.sh
RUN chmod 775 /root/run.sh

# 设置默认启动命令
CMD ["/root/run.sh"]
```

(5)查看数据库 Dockerfile 主要文件目录。

```
[root@master mysql]# tree /root/mysql
/root/mysql
├── db_init.sh
├── Dockerfile
├── gpmall.sql
├── local.repo
└── run.sh
0 directories, 5 files
```

2. Dockerfile 构建与测试

(1) docker build 命令用于使用 Dockerfile 创建镜像,因此构建 mysql 可以使用以下命令,当 Dockerfile 和当前执行命令的目录不在同一个时,也可以指定 Dockerfile。

```
[root@master mysql]# docker build -t gpmall_mysql .
Sending build context to Docker daemon  66.05kB
Step 1/16 : FROM 192.168.100.10:5000/centos:1804
 ---> cf49811e3cdb
Step 2/16 : MAINTAINER Baicai
 ---> Running in 2459ff68ff20
Removing intermediate container 2459ff68ff20
 ---> 54a9f0de7d8a
Step 3/16 : RUN rm -vf /etc/yum.repos.d/*
 ---> Running in 15fe1bdc3d8b
Removing intermediate container 15fe1bdc3d8b
 ---> f27730eba1cf
Step 4/16 : ADD local.repo /etc/yum.repos.d
 ---> 0cbbe37649dc
Step 5/16 : RUN yum -y install mariadb-server
 ---> Running in 7593c21f4988
Removing intermediate container 7593c21f4988
 ---> 0d03f9685fd5
Step 6/16 : ENV MARIADB_USER root
 ---> Running in 68c3e5ba07a7
Removing intermediate container 68c3e5ba07a7
 ---> 437fb2e07074
Step 7/16 : ENV MARIADB_PASS 123456
 ---> Running in 4fc17a8265e0
Removing intermediate container 4fc17a8265e0
 ---> 7f23e8fe810b
Step 8/16 : ENV LC_ALL en_US.UTF-8
 ---> Running in 86de59e2ca46
```

```
Removing intermediate container 86de59e2ca46
 ---> 873fed4bc620
Step 9/16 : COPY gpmall.sql /root/
 ---> 8b0c21ed9a96
Step 10/16 : ADD db_init.sh /root/db_init.sh
 ---> fdc018891da2
Step 11/16 : RUN chmod 775 /root/db_init.sh
 ---> Running in 8b6830964ab9
Removing intermediate container 8b6830964ab9
 ---> 96becad0b3da
Step 12/16 : RUN /root/db_init.sh
 ---> Running in fc4f1f7f5df3
Removing intermediate container fc4f1f7f5df3
 ---> 3d3714a62115
Step 13/16 : EXPOSE 3306
 ---> Running in e8af1e241de5
Removing intermediate container e8af1e241de5
 ---> 70ad118c7591
Step 14/16 : ADD run.sh /root/run.sh
 ---> 51353d228159
Step 15/16 : RUN chmod 775 /root/run.sh
 ---> Running in a9325d0efaf5
Removing intermediate container a9325d0efaf5
 ---> 73122962a847
Step 16/16 : CMD [ "/root/run.sh" ]
 ---> Running in b5fbe00c2fd2
Removing intermediate container b5fbe00c2fd2
 ---> b6c98485819c
Successfully built b6c98485819c
Successfully tagged gpmall_mysql:latest
```

执行命令之后，会看到控制台逐层输出构建内容，直到输出两个 Successfully 即为构建成功。

（2）镜像构建完成之后，使用 docker images 命令查看镜像列表。

```
[root@master mysql]# docker images
REPOSITORY        TAG        IMAGE ID        CREATED         SIZE
gpmall_mysql      latest     b6c98485819c    2 minutes ago   839MB
```

（3）使用该镜像启动容器进行测试，13306 是主机端口，冒号后面的 3306 是容器内部端口，并使用命令查看容器是否启动命令如下所示：

```
[root@master mysql]# docker run -itd --name mysql -p 13306:3306 gpmall_mysql
[root@master mysql]# docker ps -a
CONTAINER ID    IMAGE           COMMAND         CREATED         STATUS          PORTS               NAMES
cd374276cd1f    gpmall_mysql    "/root/run.sh"  5 seconds ago   Up 4 seconds    0.0.0.0:13306->3306/tcp   mysql
```

（4）使用远程访问，在 slave 节点安装 MariaDB-client 客户端，然后输入远程访问地址信息可以进行远程数据库的访问。命令和结果如图 8-8 所示。

```
[root@slave ~]# yum install -y MariaDB-client
```

图 8-8 远程访问数据库

本任务到此结束。

8.4 Docker 容器编排

学习目标

学完本节后，您应能够：
- 了解容器编排基础知识。
- 了解容器编排模块参数。
- 掌握容器编排案例编写。

8.4.1 Docker 容器编排简介

1. 容器编排的概念

容器编排就是指对多个容器进单独组件和应用层工作进行组织的流程。容器编排工具提供了有用且功能强大的解决方案，用于跨多个主机协调创建、管理和更新多个容器。最重要的是，业务流程允许异步地在服务和流程任务之间共享数据。在生产环境中，可以在多个服务器上运行每个服务的多个实例，以使应用程序具有高可用性。越简化编排，就越能深入了解应用程序并分解更小的微服务。我们可以利用容器编排工具对容器进行更好的管理。

2. Docker-Compose 简介

Docker-Compose 项目是 Docker 官方的开源项目，负责实现对 Docker 容器集群的快速编排。

Docker-Compose 将所管理的容器分为三层，分别是工程（project）、服务（service）以及容器（container）。Docker-Compose 运行目录下的所有文件（docker-compose.yml、extends 文件或环境变量文件等）组成一个工程，若无特殊指定工程名即为当前目录名。一个工程当中可包含多个服务，每个服务中定义了容器运行的镜像、参数、依赖。一个服务当中可包括多个容器实例，Docker-Compose 并没有解决负载均衡的问题，因此需要借助其他工具实现服务发现及负载均衡。

Docker-Compose 的工程配置文件默认为 docker-compose.yml，可通过环境变量 COMPOSE_FILE 或 -f 参数自定义配置文件，其定义了多个有依赖关系的服务及每个服务运行的容器。

使用一个 Dockerfile 模板文件，可以让用户很方便地定义一个单独的应用容器。在工作中，经常会碰到需要多个容器相互配合来完成某项任务的情况。例如要实现一个 Web 项目，除了

Web 服务容器本身，往往还需要再加上后端的数据库服务容器，甚至还包括负载均衡容器等。

Compose 允许用户通过一个单独的 docker-compose.yml 模板文件（YAML 格式）来定义一组相关联的应用容器为一个项目（project）。

Docker-Compose 项目由 Python 编写，调用 Docker 服务提供的 API 来对容器进行管理。因此，只要所操作的平台支持 Docker API，就可以在其上利用 Compose 来进行编排管理。

3. Docker Compose 调用过程

Docker-Compose 调用过程如图 8-9 所示。

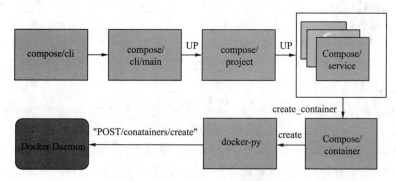

图 8-9 Docker-Compose 调用过程

（1）用户执行的 docker-compose up 指令调用了命令行中的启动方法。功能很简单明了，一个 docker-compose.yml 定义了一个 docker-compose 的 project，docker-compose 操作提供的命令行参数则作为这个 project 的启动参数，交由 project 模块去处理。

（2）如果当前宿主机已经存在与该应用对应的容器，docker-compose 将进行行为逻辑判断。如果用户指定可以重新启动已有服务，docker-compose 就会执行 service 模块的容器重启方法，否则就将直接启动已有容器。这两种操作的区别在于前者会停止旧的容器，创建启动新的容器，并把旧容器移除掉。在这个过程中创建容器的各项定义参数都是从 docker-compose up 指令和 docker-compose.yml 中传入的。

（3）启动容器的方法也很简洁，这个方法中完成了一个 Docker 容器启动所需的主要参数的封装，并在 container 模块执行启动。

（4）container 模块会调用 docker-py 客户端，执行向 Docker Daemon 发起创建容器的 POST 请求，再往后就是 Docker 处理的范畴了。

8.4.2 容器编排管理

1. docker-compose.yml 文件

Yml 是一种简洁的非标记语言。Yaml 以数据为中心，使用空白、缩进、分行组织数据，从而使得表示更加简洁易读。docker-compose.yml 常用的配置项有以下几点：

（1）build：指定 Dockerfile 所在文件夹的路径。Compose 将会利用它自动构建这个镜像，然后使用这个镜像。

```
build: /path/to/build/dir
```

（2）image：从指定的镜像中启动容器，可以是存储仓库、标签以及镜像 ID，如果镜像不存在，Compose 会自动拉取镜像。

（3）ports：定义端口映射，可以使用 HOST:CONTAINER 的方式指定端口，也可以指定容器端口（选择临时主机端口），宿主机会随机映射端口。

```
ports:
    - 83:80
    - 1443:443
```

（4）depends_on：定义依赖关系。此定义会让当前服务处于等待状态，直到这些依赖服务启动。例如某个服务依赖数据库服务，那么通过此配置解决了服务启动顺序的问题。

```
depends_on:
    - mall-mysql
    - mall-redis
    - mall-kafka
    - mall-zookeeper
```

（5）links：链接到其他服务中的容器。使用服务名称（同时作为别名）或服务名称：服务别名（SERVICE:ALIAS）格式都可以。

```
links:
    - mall-zookeeper:zookeeper.mall
```

（6）volumes：挂载一个目录或者一个已存在的数据卷容器，可以直接使用 HOST:CONTAINER 这样的格式，或者使用 HOST:CONTAINER:RO 这样的格式。

```
volumes:
- /var/lib/mysql
```

（7）args：指定构建参数，这些参数只能在构建过程中访问。

（8）command：使用 command 命令可以覆盖容器启动后默认执行的命令。

（9）dockerfile：使用此 dockerfile 文件来构建，必须指定构建路径。

2. docker-compose.yaml 解读

如下是 docker-compose.yaml 的编排文件：

```
[root@master gpmall-compose]# vi docker-compose.yaml
version: '3.0'                              #docker对应当前compose版本号
services:                                   # 容器组
  mall-kafka:                               # 镜像容器服务标识
    image: gpmall_kafka                     # 使用的镜像名称
    container_name: mall-kafka              # 容器名称
    hostname: kafka.mall                    # 容器内主机名称
    ports:                                  # 容器的映射端口
      - 19092:9092                          # 端口号
    depends_on:                             # 指定服务之间的依赖关系
      - mall-zookeeper                      # 依赖的服务容器名称
    links:                                  # 链接容器
      - mall-zookeeper:zookeeper.mall       # 容器的链接关系
```

执行 yml 文件中 image 时，首先拉取本地镜像文件，基于此镜像启动容器。

8.4.3 任务：docker-compose 编排

docker compose 在 Docker 容器运用中具有重要的学习意义，docker compose 是一个整合发布应用的利器。而使用 docker compose 时，懂得如何编排 docker compose 配置文件也是很重要的。

任务执行清单

在本任务中，您将通过实践编写 docker-compose.yaml 文件一键编排 owncloud 服务。

第 8 章 容器高级运维

☕ **目标**
- 掌握 docker-compose 编排参数。
- 掌握 docker-compose 编排服务。

📎 **重要信息**
- 本任务采用 CentOS 7.5_1804 操作系统。
- 虚拟机密码为 000000。
- 提前配置好 yum 源仓库。
- 关闭防火墙和 selinux。
- 所有节点已安装好 docker-ce。

✏️ **解决方案**

1. 编排 owncloud

（1）登录到 master 节点虚拟机，将镜像中的 docker-compose 命令启动配置文件复制到 /usr/loca/bin/ 目录下，命令和结果如下所示：

```
[root@master ~]# cp -rvf /opt/docker/docker-compose/v1.25.5-docker-compose-Linux-x86_64  /usr/local/bin/docker-compose
'/opt/docker/docker-compose/v1.25.5-docker-compose-Linux-x86_64' -> '/usr/local/bin/docker-compose'
```

（2）登录到 master 节点虚拟机，更改 docker-compose 文件的执行权限并查看版本信息，命令和结果如下所示：

```
[root@master ~]# chmod +x /usr/local/bin/docker-compose
[root@master ~]# docker-compose version
docker-compose version 1.25.5, build 8a1c60f6
docker-py version: 4.1.0
CPython version: 3.7.5
OpenSSL version: OpenSSL 1.1.0l  10 Sep 2019
```

（3）服务配置完成之后，先将准备好的 owncloud 和 mysql 镜像导入到本地仓库中，命令如下所示：

```
[root@master images]# docker load -i owncloud-latest.tar
[root@master images]# docker load -i mysql-5.6.tar
```

（4）创建 owncloud 文件夹，编写 owncloud 的 docker-compose 配置文件，详细信息如下所示：

```
[root@master ~]# mkdir owncloud
[root@master ~]# cd owncloud/
[root@master owncloud]# vi docker-compose.yaml
version: '3.0'
services:
  owncloud:
    image: owncloud:latest
    container_name: owncloud
    volumes:
      - /data/db/owncloud:/var/www/html/data
    ports:
      - 5679:80
  mysql:
    image: mysql:5.6
```

```
    container_name: owncloud-db
    environment:
      - "MYSQL_ROOT_PASSWORD=123456"
    ports:
      - 3306:3306
```

2. 启动和检测

(1) docker-compose 配置文件编写好之后,使用 docker-compose up 命令一键启动容器,然后查看容器的详细信息,命令和结果如下所示:

```
[root@master owncloud]# docker-compose up -d
Creating network "owncloud_default" with the default driver
Creating owncloud-db ... done
Creating owncloud    ... done
[root@master owncloud]# docker-compose ps
    Name            Command              State          Ports
-------------------------------------------------------------------------
owncloud        docker-entrypoint.sh apach ...   Up      0.0.0.0:5679->80/tcp
owncloud-db     docker-entrypoint.sh mysqld      Up      0.0.0.0:3306->3306/tcp
```

(2) 在浏览器中输入 http://IP:5679 即可看见图 8-10 所示的界面,docker-compose 部署 owncloud 成功。

图 8-10 owncloud 登录界面

(3) 在平台中填入参数信息,即可输入 admin:admin 进行登录,结果如图 8-11 所示。

```
用户名: admin
密码: admin
数据库用户: root
数据库密码: 123456
数据库名: owncloud
服务器地址: 192.168.100.10
```

图 8-11 owncloud 服务网站

本任务到此结束。

8.5 开放研究任务：容器应用商城系统编排

某公司将要开发一套 Web 应用系统 gpmall 应用商城，实现全容器化部署基于服务器的性能、可靠性、高可用性与方便维护，公司研发部决定使用微服务架构开发的 Web 应用系统产品。

任务执行清单

在本任务中，您将通过编写多例 Dockerfile 文件来构建镜像，使用 docker-compose 将镜像进行编排，完成 gpmall 应用商城的部署。

（1）基础环境配置。
（2）Redis 服务构建。
（3）Zookeeper 服务构建。
（4）Kafka 服务构建。
（5）Nginx 服务构建。
（6）docker-compose 编排。

目标

- 掌握 Dockerfile 编写。
- 掌握 docker-compose 编排。
- 掌握 gpmall 应用商城部署。

重要信息

- 本任务采用 CentOS 7 操作系统。
- 虚拟机密码为 000000。
- 提前配置好 yum 源仓库。
- 关闭防火墙和 Selinux。
- 所有节点已安装好 docker-ce。

解决方案

1. 基础环境配置

（1）登录到 master 节点，更改该节点的 yum 源文件，都设置为 ftp 服务模式，命令和结果如下所示：

```
[root@master ~]# vi /etc/yum.repos.d/gpmall.repo
[docker]
name=docker
baseurl=ftp://192.168.100.10/docker/kubernetes-repo
enabled=1
gpgcheck=0
[centos]
name=centos
baseurl=ftp://192.168.100.10/centos
enabled=1
gpgcheck=0
[gpmall]
name=gpmall
```

```
baseurl=ftp://192.168.100.10/docker/ChinaskillMall/gpmall-repo
enabled=1
gpgcheck=0
```

（2）由于 mysql 镜像在前面章节已经编写完成，此处不再赘述，后期可直接使用。使用命令查看镜像列表，结果如下所示：

```
[root@master ~]# docker images
REPOSITORY      TAG       IMAGE ID        CREATED           SIZE
gpmall_mysql    latest    b6c98485819c    1 minutes ago     839MB
```

2. Redis 服务构建

在部署 gpmall 应用商城的时候需要使用 Redis 缓存服务器来提高服务的性能。

（1）在 master 节点上编写 /root/redis/Dockerfile 文件构建 gpmall_redis 镜像。首先创建一个用于存放 Redis 服务的文件夹，然后编写 Dockerfile 文件，详细信息如下所示：

```
[root@master ~]# mkdir /root/redis
[root@master ~]# cd /root/redis/
[root@master redis]# vi Dockerfile
FROM 192.168.100.10:5000/centos:1804
MAINTAINER Baicai
RUN rm -rf /etc/yum.repos.d/*
COPY gpmall.repo /etc/yum.repos.d/
RUN yum install -y redis && \
    sed -i 's/^bind 127.0.0.1/bind 0.0.0.0/g' /etc/redis.conf && \
    sed -i 's/^protected-mode yes/protected-mode no/g' /etc/redis.conf
EXPOSE 6379
ENTRYPOINT ["/usr/bin/redis-server","/etc/redis.conf"]
```

（2）将 yum 源文件 gpmall.repo 复制到当前目录下，然后使用 tree 命令查看当前目录结构，命令和结果如下所示：

```
[root@master redis]# cp /etc/yum.repos.d/gpmall.repo .
[root@master redis]# tree /root/redis
/root/redis
├── Dockerfile
└── gpmall.repo
0 directories, 2 files
```

（3）使用 docker build 命令构建 gpmall_redis 镜像，命令和结果如下所示：

```
[root@master redis]# docker build -t gpmall_redis
Sending build context to Docker daemon  3.072kB
Step 1/7 : FROM 192.168.100.10:5000/centos:1804
 ---> cf49811e3cdb
Step 2/7 : MAINTAINER Baicai
 ---> Using cache
 ---> 54a9f0de7d8a
Step 3/7 : RUN rm -rf /etc/yum.repos.d/*
 ---> Running in 5a09d826f116
Removing intermediate container 5a09d826f116
 ---> 4f528323cc4f
Step 4/7 : COPY gpmall.repo /etc/yum.repos.d/
 ---> 3d9c3f82a662
Step 5/7 : RUN yum install -y redis && sed -i 's/^bind 127.0.0.1/bind 0.0.0.0/g' /etc/redis.conf && sed -i 's/^protected-mode yes/protected-mode no/g' /etc/redis.conf
```

```
---> Running in e1ebc6bde4cb
Removing intermediate container e1ebc6bde4cb
 ---> 60c79f35c7dd
Step 6/7 : EXPOSE 6379
 ---> Running in e37d68cd3568
Removing intermediate container e37d68cd3568
 ---> a5a5c2a22990
Step 7/7 : ENTRYPOINT ["/usr/bin/redis-server","/etc/redis.conf"]
 ---> Running in 8df864dd2a1a
Removing intermediate container 8df864dd2a1a
 ---> fae21cd643e7
Successfully built fae21cd643e7
Successfully tagged gpmall_redis:latest
```

（4）镜像构建成功之后，在 master 节点安装 Redis 服务，然后使用该镜像指定 6379 端口运行 redis 容器，命令和结果如下所示：

```
[root@master redis]# yum install -y redis
[root@master redis]# docker run -itd --name redis -p 6379:6379 gpmall_redis
dcb87a2a2df477604c3006e22265885ce5e5e6ebe582a365756b3131c7eda7b0
```

（5）使用 redis 命令指定 6379 端口，进行 Redis 服务的访问，测试成功删除容器，命令和测试结果如下所示：

```
[root@master redis]# redis-cli -p 6379
127.0.0.1:6379> ping
PONG
[root@master redis]# docker rm -f redis
redis
```

3. Zookeeper 服务构建

（1）在 master 节点上编写 /root/zookeeper/Dockerfile 文件，构建 gpmall_zookeeper 镜像。首先创建一个用于存放 Zookeeper 服务的文件夹，然后将 zookeeper-3.4.14.tar.gz 安装包上传至该文件夹下。

```
[root@master ~]# mkdir /root/zookeeper
[root@master ~]# cd /root/zookeeper/
[root@master zookeeper]# ls
zookeeper-3.4.14.tar.gz gpmall.repo
```

（2）编写 Dockerfile 文件，再将 gpmall.repo 文件上传至该目录，编写完成之后使用 tree 命令查看目录结构，详细信息如下所示：

```
[root@master zookeeper]# vi Dockerfile
FROM 192.168.100.10:5000/centos:1804
MAINTAINER Baicai
RUN rm -rf /etc/yum.repos.d/*
COPY gpmall.repo /etc/yum.repos.d/
ADD zookeeper-3.4.14.tar.gz /root
RUN yum install -y java java-devel && \
    mv /root/zookeeper-3.4.14/conf/zoo_sample.cfg /root/zookeeper-3.4.14/conf/zoo.cfg
EXPOSE 2181
ENTRYPOINT ["/root/zookeeper-3.4.14/bin/zkServer.sh","start-foreground"]
[root@master zookeeper]# tree /root/zookeeper
/root/zookeeper
```

```
├── Dockerfile
├── gpmall.repo
└── zookeeper-3.4.14.tar.gz
0 directories, 3 files
```

（3）使用 docker build 命令构建 gpmall_zookeeper 镜像，命令和结果如下所示：

```
[root@master zookeeper]# docker build -t gpmall_zookeeper
Sending build context to Docker daemon  37.68MB
Step 1/8 : FROM 192.168.100.10:5000/centos:1804
 ---> cf49811e3cdb
Step 2/8 : MAINTAINER Baicai
 ---> Using cache
 ---> 54a9f0de7d8a
Step 3/8 : RUN rm -rf /etc/yum.repos.d/*
 ---> Using cache
 ---> 4f528323cc4f
Step 4/8 : COPY gpmall.repo /etc/yum.repos.d/
 ---> Using cache
 ---> 3d9c3f82a662
Step 5/8 : ADD zookeeper-3.4.14.tar.gz /root
 ---> b5a04372421f
Step 6/8 : RUN yum install -y java java-devel &&    mv /root/zookeeper-3.4.14/conf/zoo_sample.cfg /root/zookeeper-3.4.14/conf/zoo.cfg
 ---> Running in a002593c4a2a
Removing intermediate container a002593c4a2a
 ---> 98f5a8af461a
Step 7/8 : EXPOSE 2181
 ---> Running in e2a06ae994dd
Removing intermediate container e2a06ae994dd
 ---> 5b8403baae13
Step 8/8 : ENTRYPOINT ["/root/zookeeper-3.4.14/bin/zkServer.sh","start-foreground"]
 ---> Running in ec875251c372
Removing intermediate container ec875251c372
 ---> 3af8cd707ebb
Successfully built 3af8cd707ebb
Successfully tagged gpmall_zookeeper:latest
```

（4）镜像构建成功之后，在 master 节点通过刚构建的 gpmall_zookeeper 镜像创建 zookeeper 容器，使用 docker-exec 命令进入该容器查看 Zookeeper 服务的状态信息。

```
[root@master ~]# docker run -itd --name zookeeper -p 2181:2181 gpmall_zookeeper
9d99b6cf7f3d0d8606fd79f817229a06e283998e731f99df7cec632ac270aa43
[root@master ~]# docker exec -it zookeeper /bin/bash
[root@9d99b6cf7f3d /]# /root/zookeeper-3.4.14/bin/zkServer.sh status
ZooKeeper JMX enabled by default
Using config: /root/zookeeper-3.4.14/bin/../conf/zoo.cfg
Mode: standalone
[root@9d99b6cf7f3d /]# exit
exit
```

4．Kafka 服务构建

（1）首先创建一个用于存放 Kafka 服务的文件夹，将 yum 源文件和 kafka_2.11-1.1.1.tgz 压缩文件移动和上传至该目录下。

```
[root@master ~]# mkdir /root/kafka
[root@master ~]# cd /root/kafka/
[root@master kafka]# ls
gpmall.repo  kafka_2.11-1.1.1.tgz
```

（2）编写 /root/kafka/Dockerfile 文件，构建 gpmall_kafka 镜像，详细文件内容如下所示：

```
[root@master kafka]# vi Dockerfile
FROM 192.168.100.10:5000/centos:1804
MAINTAINER Baicai
RUN rm -rf /etc/yum.repos.d/*
COPY gpmall.repo /etc/yum.repos.d/
ADD kafka_2.11-1.1.1.tgz /root
RUN yum install -y java java-devel && \
    sed -i 's/^zookeeper.connect=localhost:2181/zookeeper.connect=zookeeper.mall:2181/g' /root/kafka_2.11-1.1.1/config/server.properties
EXPOSE 9092
ENTRYPOINT ["/root/kafka_2.11-1.1.1/bin/kafka-server-start.sh","/root/kafka_2.11-1.1.1/config/server.properties"]
[root@master kafka]# tree /root/kafka
/root/kafka
├── Dockerfile
├── gpmall.repo
└── kafka_2.11-1.1.1.tgz
0 directories, 3 files
```

（3）使用 docker build 命令构建 gpmall_kafka 镜像，命令和结果如下所示：

```
[root@master kafka]# docker build -t gpmall_kafka
Sending build context to Docker daemon  57.48MB
Step 1/8 : FROM 192.168.100.10:5000/centos:1804
 ---> cf49811e3cdb
Step 2/8 : MAINTAINER Baicai
 ---> Using cache
 ---> 54a9f0de7d8a
Step 3/8 : RUN rm -rf /etc/yum.repos.d/*
 ---> Using cache
 ---> 4f528323cc4f
Step 4/8 : COPY gpmall.repo /etc/yum.repos.d/
 ---> Using cache
 ---> 3d9c3f82a662
Step 5/8 : ADD kafka_2.11-1.1.1.tgz /root
 ---> a3aea9413257
Step 6/8 : RUN yum install -y java java-devel &&  sed -i 's/^zookeeper.connect=localhost:2181/zookeeper.connect=zookeeper.mall:2181/g' /root/kafka_2.11-1.1.1/config/server.properties
 ---> Running in c377fdeb7a25
Removing intermediate container c377fdeb7a25
 ---> 4c03b6679a24
Step 7/8 : EXPOSE 9092
 ---> Running in 64d8f10dae62
Removing intermediate container 64d8f10dae62
 ---> 4f8450cde8c8
Step 8/8 : ENTRYPOINT ["/root/kafka_2.11-1.1.1/bin/kafka-server-start.sh","/root/kafka_2.11-1.1.1/config/server.properties"]
```

```
---> Running in bcee4646dded
Removing intermediate container bcee4646dded
 ---> b3d7c3b42e78
Successfully built b3d7c3b42e78
Successfully tagged gpmall_kafka:latest
```

（4）镜像构建完成，下面进行镜像测试，首先查看上一步创建的 zookeeper 容器的 IP 地址，用于映射 Kafka 服务启动，命令和结果如下所示：

```
[root@master kafka]# docker inspect zookeeper  | grep IPAddress
            "IPAddress": "172.17.0.4",
```

（5）启动 kafka 容器，在容器中增加一个主机映射，信息为 Zookeeper 容器，因为 kafka 服务依赖于 Zookeeper 服务，容器启动之后进入 kafka 容器查看服务的状态信息，最后退出容器并删除，命令和结果如下所示：

```
[root@master kafka]# docker run -itd --name kafka -p 9092:9092 --add-host
zookeeper.mall:172.17.0.4 gpmall_kafka
7ed319252d666f0119af17dae0db56cb1ff79d7b0174cbcf472f47f0a9d91d12
[root@master kafka]# docker exec -it kafka /bin/bash
[root@7ed319252d66 /]# jps
352 Jps
1 Kafka
[root@7ed319252d66 /]# exit
exit
[root@master kafka]# docker rm -f zookeeper kafka
zookeeper
kafka
```

5. Nginx 服务构建

（1）首先创建一个用于存放 Nginx 服务的文件夹，然后将 Java 开发包环境、dist 压缩包和 gpmall.repo 文件上传至该文件夹下，操作如下所示：

```
[root@master ~]# mkdir nginx
[root@master ~]# cd nginx/
[root@master nginx]# ll
total 181780
-rw-r--r--. 1 root root  2501567 Apr 24 14:20    dist.tar.gz
-rw-r--r--. 1 root root      277 Apr 24 14:22    gpmall.repo
-rw-r--r--. 1 root root 39005468 Apr 24 14:09    gpmall-user-0.0.1-
SNAPSHOT.jar
-rw-r--r--. 1 root root 62386947 Apr 24 14:09    user-provider-0.0.1-
SNAPSHOT.jar
-rw-r--r--. 1 root root 47765224 Apr 24 14:09    gpmall-shopping-0.0.1-
SNAPSHOT.jar
-rw-r--r--. 1 root root 54936064 Apr 24 14:09    shopping-provider-0.0.1-
SNAPSHOT.jar
```

（2）先在 master 节点安装 Nginx 服务，然后进入 Nginx 的默认配置文件进行配置，将配置好的文件复制到 /root/nginx 目录下，配置信息如下所示：

```
[root@master nginx]# yum install -y nginx
[root@master nginx]# vi /etc/nginx/conf.d/default.conf
...
    location /user {
```

```
        proxy_pass http://127.0.0.1:8082;
    }
    location /shopping {
        proxy_pass http://127.0.0.1:8081;
    }
    location /cashier {
        proxy_pass http://127.0.0.1:8083;
    }
[root@master nginx]# cp /etc/nginx/conf.d/default.conf .
```

（3）因为 gpmall 是运行在 Nginx 服务器中，所以在使用 Nginx 镜像启动容器的时候是已经配置好所有信息的，此处需要编写 start.sh 脚本文件，用于在启动 Nginx 容器后跑完 Java 程序，详细信息如下所示：

```
[root@master nginx]# vi start.sh
#!/bin/bash
/usr/sbin/nginx -c /etc/nginx/nginx.conf &
java -jar /root/user-provider-0.0.1-SNAPSHOT.jar &
sleep 20
java -jar /root/shopping-provider-0.0.1-SNAPSHOT.jar &
sleep 20
java -jar /root/gpmall-user-0.0.1-SNAPSHOT.jar &
sleep 20
java -jar /root/gpmall-shopping-0.0.1-SNAPSHOT.jar
```

（4）编写 Nginx 服务的 Dockerfile 文件，用于构建 gpmall_nginx 镜像，详细信息如下所示：

```
[root@master nginx]# vi Dockerfile
FROM 192.168.100.10:5000/centos:1804
MAINTAINER Baicai
RUN rm -rf /etc/yum.repos.d/*
COPY gpmall.repo /etc/yum.repos.d/
COPY *.jar /root/
COPY start.sh /root/
COPY default.conf /root
ADD dist.tar.gz /root/
RUN yum install -y java java-devel nginx && \
    mv /root/dist/* /usr/share/nginx/html && \
    rm -rf /etc/nginx/conf.d/default.conf && \
    mv /root/default.conf /etc/nginx/conf.d && \
    chmod +x /root/start.sh
EXPOSE 80
EXPOSE 443
EXPOSE 8081
EXPOSE 8082
EXPOSE 8083
ENTRYPOINT ["/root/start.sh",""]
```

（5）使用 docker build 命令构建 gpmall_nginx 镜像，过程和结果如下所示：

```
[root@master nginx]# docker build -t gpmall_nginx .
Sending build context to Docker daemon  218.1MB
Step 1/15 : FROM 192.168.100.10:5000/centos:1804
 ---> cf49811e3cdb
Step 2/15 : MAINTAINER Baicai
 ---> Using cache
```

```
 ---> 54a9f0de7d8a
Step 3/15 : RUN rm -rf /etc/yum.repos.d/*
 ---> Using cache
 ---> 4f528323cc4f
Step 4/15 : COPY gpmall.repo /etc/yum.repos.d/
 ---> Using cache
 ---> 3d9c3f82a662
Step 5/15 : COPY *.jar /root/
 ---> Using cache
 ---> 4433e8674daf
Step 6/15 : COPY start.sh /root/
 ---> Using cache
 ---> 7fafad681b27
Step 7/15 : COPY default.conf /root
 ---> Using cache
 ---> 68f709941708
Step 8/15 : ADD dist.tar.gz /root/
 ---> Using cache
 ---> ae8f6d1a7202
Step 9/15 : RUN yum install -y java java-devel nginx &&    mv /root/dist/* /usr/share/nginx/html &&    rm -rf /etc/nginx/conf.d/default.conf && mv /root/default.conf /etc/nginx/conf.d &&    chmod +x /root/start.sh
 ---> Using cache
 ---> d917940bf67e
Step 10/15 : EXPOSE 80
 ---> Using cache
 ---> 57ced3db6964
Step 11/15 : EXPOSE 443
 ---> Using cache
 ---> fdc60b638731
Step 12/15 : EXPOSE 8081
 ---> Using cache
 ---> 4ef130c5f105
Step 13/15 : EXPOSE 8082
 ---> Using cache
 ---> 1926aac137a1
Step 14/15 : EXPOSE 8083
 ---> Using cache
 ---> e0ef916b208f
Step 15/15 : ENTRYPOINT ["/root/start.sh",""]
 ---> Using cache
 ---> 03518302c2e4
Successfully built 03518302c2e4
Successfully tagged gpmall_nginx:latest
```

执行命令之后，会看到控制台逐层输出构建内容，直到输出两个 Successfully 即为构建成功。

（6）使用 docker-images 查看镜像列表，命令和结果如图 8-12 所示。

```
[root@master `]# docker images
REPOSITORY            TAG         IMAGE ID          CREATED              SIZE
gpmall_nginx          latest      03518302c2e4      16 minutes ago       714MB
gpmall_kafka          latest      509ee7139dd9      About an hour ago    542MB
gpmall_zookeeper      latest      3af8cd707ebb      About an hour ago    540MB
gpmall_redis          latest      fae21cd643e7      2 hours ago          247MB
gpmall_mysql          latest      b6c98485819c      5 hours ago          839MB
```

图 8-12　镜像列表

6. docker-compose 编排

(1) 创建 docker-compose 专属文件夹,名称为 gpmall-compose,根据实验要求编写的 gpmall-compose.yaml 详细内容如下所示,在编写过程中要注意文件的格式信息,如有错误进行调试修改即可。

```
[root@master ~]# mkdir /root/gpmall-compose
[root@master ~]# cd gpmall-compose/
[root@master gpmall-compose]# vi docker-compose.yaml
version: '3.0'
services:
  mall-mysql:
    image: gpmall_mysql
    container_name: mall-mysql
    hostname: mysql.mall
    ports:
      - 13306:3306
  mall-redis:
    image: gpmall_redis
    container_name: mall-redis
    hostname: redis.mall
    ports:
      - 16379:6379
  mall-kafka:
    image: gpmall_kafka
    container_name: mall-kafka
    hostname: kafka.mall
    ports:
      - 19092:9092
    depends_on:
      - mall-zookeeper
    links:
      - mall-zookeeper:zookeeper.mall
  mall-zookeeper:
    image: gpmall_zookeeper
    container_name: mall-zookeeper
    hostname: zookeeper.mall
    ports:
      - 12181:2181
  mall-nginx:
    image: gpmall_nginx
    container_name: mall-nginx
    hostname: nginx.mall
    ports:
      - 83:80
      - 1443:443
    depends_on:
      - mall-mysql
      - mall-redis
      - mall-kafka
      - mall-zookeeper
    links:
      - mall-mysql:mysql.mall
      - mall-redis:redis.mall
      - mall-kafka:kafka.mall
      - mall-zookeeper:zookeeper.mall
```

(2) 使用 docker-compose 命令在后台启动所有容器,使用 docker-compose ps 命令查看

compose 容器列表详细信息,其中包含了容器名、端口映射、状态等,命令和结果如下所示:

```
[root@master gpmall-compose]# docker-compose up -d
Creating mall-mysql      ... done
Creating mall-redis      ... done
Creating mall-zookeeper  ... done
Creating mall-kafka      ... done
Creating mall-nginx      ... done
[root@master gpmall-compose]# docker-compose ps
   Name              Command                    State           Ports
-----------------------------------------------------------------------------
 mall-kafka      /root/kafka_2.11-1.1.1/bin ...  Up    0.0.0.0:19092->9092/tcp
 mall-mysql      /root/run.sh                    Up    0.0.0.0:13306->3306/tcp
 mall-nginx      /root/start.sh                  Up    0.0.0.0:1443->443/tcp, 0.0.0.0:83->80/tcp, 8081/tcp, 8082/tcp, 8083/tcp
 mall-redis      /usr/bin/redis-server /etc ...  Up    0.0.0.0:16379->6379/tcp
 mall-zookeeper  /root/zookeeper-3.4.14/bin ...  Up    0.0.0.0:12181->2181/tcp
```

(3) 大约五分钟后,等待 Nginx 容器中的 Java 程序跑完,在浏览器中输入 http://192.168.100.10:83,效果如图 8-13 所示。至此,使用容器部署 gpmall 应用商城系统成功。

图 8-13 gpmall 应用商城系统

本任务到此结束。

小 结

在本章中,您已经学会:
- Docker 容器技术的原理和架构。
- Docker 容器的安装与仓库配置。
- Docker 容器的基础管理。
- Dockerfile 编写。
- Docker 容器编排。
- 容器应用商城系统编排。

第 9 章

Kubernetes 服务架构与平台

本章概要

学习在 Linux 操作系统环境下,通过部署 Kubernetes 服务,基于 Kubernetes 编排服务,完成基于 Kubernetes 服务编排部署 WordPress+MySQL 案例部署。

学习目标

- 学习 Kubernetes 服务的原理和架构。
- 掌握 Kubernetes 集群架构的部署。
- 掌握 Kubernetes 编排案例的部署。
- 掌握 Kubernetes 编排 WordPress。

思维导图

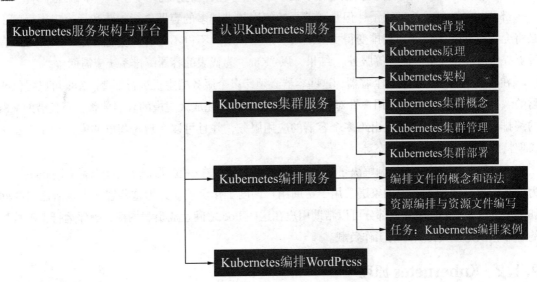

任务目标

学习 Kubernetes 集群服务的部署与管理,通过学习相应模块,完成 Kubernetes 编排 WordPress 案例。

9.1 认识 Kubernetes 服务

学习目标

学完本节后，您应能够：
- 了解 Kubernetes 背景。
- 了解 Kubernetes 原理。
- 了解 Kubernetes 架构。

9.1.1 Kubernetes 背景

1. Kubernets 简介

Kubernetes 简称 K8s（以 K 开头，以 S 结尾，中间 8 个字母以数字 8 的形式表达）。Kubernetes 官网：https://.kubernetes.io。Kubernetes 是一个开源的系统，它主要用于自动部署，扩容缩容和管理容器应用。Kubernetes 将诸多应用的容器分为若干个逻辑单元以便于管理。Kubernetes 拥有着 Google 高负载生产环境的十余年经验，它起源于 Google 内部平台的一个框架——Borg。Kubernetes 项目来源于 Borg，可以说是集结了 Borg 设计思想的精华，并且吸收了 Borg 系统中的经验和教训。

2. Kubernets 背景

近 20 年，IT 领域的新技术、新概念层出不穷，如 devops、微服务、容器、云计算、区块链等。以 Docker 为首的容器化技术解决了 devops 中"因为 环境、配置及程序本身不同 而造成的各种部署配置的问题"，Docker 将它们统一在容器镜像之上；如今，越来越多的企业选择以"镜像文件作为交付载体"，镜像内直接包括了程序本身及其依赖的系统环境、库、基础程序等。

部署的复杂度虽然解决了，但是随着开始使用越来越多的容器进行封装和运行应用程序，必将会导致容器的管理和编排变得更加困难。最终，用户不得不对容器实施分组，以便跨所有容器提供网络、安全、监控等服务。结果，以"K8s"为代表的容器编排系统应需而生。

在生产环境中会涉及多个容器，这些容器必须跨多个服务器主机进行部署。K8s 可以提供所需的编排和管理功能，以便用户针对这些工作负载轻松地完成大规模的容器部署，而且借助 K8s 的编排功能，用户可以构建出跨多个容器的应用服务，并且可以实现跨集群调度、扩展容器、长期持续管理容器的健康状态等。

K8s 利用容器的扩容机制解决了很多问题。它将容器放在一起，形成了一个"容器集（Pods）"，为分组的容器增加了一个抽象层，用于帮助用户调度工作负载，并为这些容器提供所需的联网和存储等服务，K8s 的其他部分可以帮助用户在这些 Pod 之间达成负载均衡，确保运行正确数量的容器，以充分支持实际的工作负载。

9.1.2 Kubernetes 原理

1. Kubernets 核心服务

K8s 最核心的组件是 Service（服务），Service 是由 Pod 组成的，而 Pod 是由容器组成的。提供 Service 的是容器，为了确保 Service 的高可用，提供 Service 的容器不能只有一个，而是一组，这一组容器称为 Pod。Pod 和 Service 之间通过一个叫作 Label 的标签进行关联，用 Label 可以给

Pod 打标记。

Pod 都是运行在 Node 上的，我们把物理机、虚拟机、云主机称为 Node。通常情况下一个 Node 上可以运行几百个 Pod，Pod 是由若干个容器组成的，Pod 里面有一个核心的容器叫作 Pause，Pause 和普通的容器不太一样，它拥有一个很重要的功能，即用来共享网络栈和 Volume 挂载卷，因此同一个 Pod 内的业务容器之间的通信和数据交换更为高效。

2. Kubernetes 工作原理

Kubernetes 工作流程如图 9-1 所示。

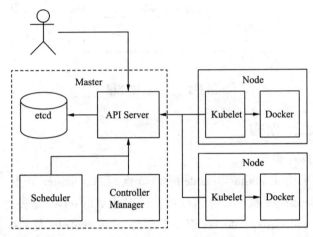

图 9-1　Kubernetes 工作流程

以下为 Pod 的创建过程：

（1）用户提交创建 Pod 的请求，可以通过 API Server 的 REST API，也可用 Kubectl 命令行工具，支持 Json 和 Yaml 两种格式。

（2）API Server 处理用户请求，存储 Pod 数据到 etcd。

（3）Schedule 通过和 API Server 的 watch 机制，查看到新的 Pod，尝试为 Pod 绑定 Node。

（4）过滤主机：调度器用一组规则过滤掉不符合要求的主机，例如 Pod 指定了所需要的资源，就要过滤掉资源不够的主机。

（5）主机打分：对第一步筛选出的符合要求的主机进行打分，在主机打分阶段，调度器会考虑一些整体优化策略，例如把一个 Replication Controller 的副本分布到不同的主机上，使用最低负载的主机等。

（6）选择主机：选择打分最高的主机，进行 binding 操作，将结果存储到 etcd 中。

（7）Kubelet 根据调度结果执行 Pod 创建操作：绑定成功后，会启动 container，docker run，Scheduler 会调用 API 在数据库 etcd 中创建一个 bound pod 对象，描述在一个工作节点上绑定运行的所有 Pod 信息。运行在每个工作节点上的 Kubelet 也会定期与 etcd 同步 bound pod 信息，一旦发现应该在该工作节点上运行的 bound pod 对象没有更新，则调用 Docker API 创建并启动 Pod 内的容器，如图 9-2 所示。

在此期间，Control Manager 会同时根据 Kubernetes 的 mainfiles 文件执行 rc pod 的数量来保证指定的 Pod 副本数。而其他的组件，例如 Scheduler 负责 Pod 绑定的调度，从而完成整个 Pod 的创建。

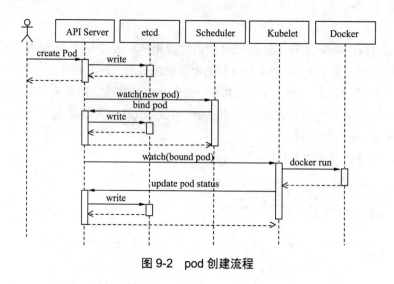

图 9-2　pod 创建流程

9.1.3　Kubernetes 架构

1. Kubernetes 服务架构

Kubernetes 集群主要由 master 和 node 两类节点组成。master 节点主要由 API server、Controller Manager、Scheduler 和 etcd 等组件组成，其中 API server 是整个集群的网关。node 节点主要由 kubelet、kube-proxy、docker 引擎等组件组成。kubelet 是 K8S 集群的工作与节点上的代理组件。一个完整的 Kubernetes 集群还包括 CoreDNS、Prometheus（或 HeapSter）、Dashboard、Ingress Controller 等附加组件。其中 cAdivsor 组件作用于各个节点（Master 和 Node 节点）之上，用于收集容器及节点的 CPU、内存以及磁盘资源的利用率指标数据，这些统计数据由 Heapster 聚合后，可以通过 API Server 访问。Kubernetes 节点有运行应用容器必备的服务，而这些都受 Master 的控制。每个节点上都要运行 Docker。Docker 负责所有具体的映像下载和容器运行。Kubernetes 的架构如图 9-3 所示。

整体来看，是一个 Master 节点，多个 Node 节点的结构，基本上所有的分布式系统都是如此。

2. Kubernetes 组件

（1）元数据存储与集群维护。

一个集群系统有一个统一的地方维护整个集群以及任务的元数据。而集群系统的控制节点，为了高可用性，往往存在多个 Master，在多个 Master 中间也有一个 Leader。

（2）API 层与命令行。作为一个分布式系统，每一层都会有自己的 API，但是对外往往需要一个统一的 API 接口层，一般除了界面之外，为了自动化，往往会有一个命令行可以执行操作，其实命令里面封装的也是对 API 的调用。

（3）调度。在运行一个容器的时候，放在哪个节点上，这个过程是调度。对于 Kubernetes，调度是由一个单独的进程负责的。

（4）副本与弹性伸缩。容器部署无状态服务的一个优点就是可以多副本，并且可以弹性伸缩。

（5）编排。为了能够通过编排文件一键创建整个应用，需要有编排功能。Kubernetes 的编排是基于 yml 文件。

（6）容器。Kubernetes 支持 Docker、RKT 等多种容器格式。

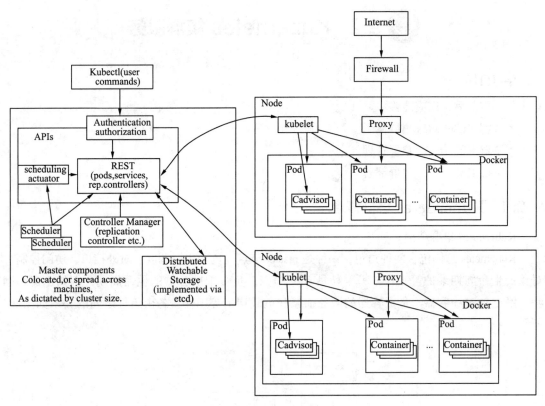

图 9-3　Kubernetes 架构

（7）网络。容器的网络配置有 Docker Libnetwork Container Network Model（CNM）和 Container Network Interface（CNI）两个阵营。Docker Libnetwork 的优势就是原生，而且和 Docker 容器生命周期结合紧密。

（8）存储。Kubernetes 可以创建 Persistent Volumes，支持 GCE、AWS、NFS、GlusterFS、Ceph 等。

（9）监控。容器的监控中 Prometheus + cadvisor 是主流的方案，而 cadvisor 来自 Kubernetes 的一个组件。

（10）节点。Kubernetes 的 Node 上运行的是 kubelet。

3．kubernetes 核心对象

Cluster：计算、存储、网络资源的集合，kubernetes 利用这些资源运行各种基于容器的应用。

Master：是 Cluster 的大脑，主要职责是调度，即决定将应用放在那里运行，可以运行多个 Master。

Node：职责是运行容器应用，Node 由 Master 管理，Node 负责监控并汇报容器的状态，bin 根据 Master 的要求管理容器的生命周期。

Pod：是 kubernetes 的最小工作单元，每个 Pod 包含一个或多个容器，Pod 中的容器被作为一个整体被 Master 调度到一个 Node 上运行，目的在于提高可管理性及通信和资源共享能力。

9.2 Kubernetes 集群服务

学习目标

学完本节后,您应能够:
- 了解 Kubernetes 集群概念。
- 了解 Kubernetes 集群管理。
- 掌握 Kubernetes 集群部署。

9.2.1 Kubernetes 集群概念

Kubernetes 集群结构

Kubernetes 总体包含两种角色,一种是 master 节点,负责集群调度、对外接口、访问控制、对象的生命周期维护等工作;另一种是 node 节点,负责维护容器的生命周期,例如创建、删除、停止 Docker 容器,负责容器的服务抽象和负载均衡等工作。各个组件之间的关系如图 9-4 所示。

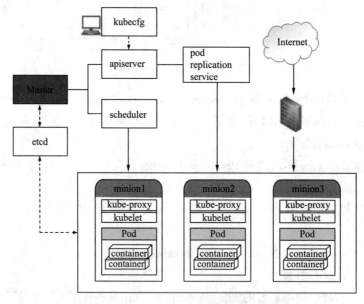

图 9-4 Kubernetes 集群服务各组件关系

1. master 节点

master 是一个管理中心,也叫作集群的控制中心,kubernetes 的所有指令都是发送给 master,它负责具体的执行过程。一般会把 master 独立于一台物理机或者一台虚拟机,它的重要性不言而喻。master 上有 4 个关键的进程:

(1) API Server:提供了资源对象的唯一操作入口,其他所有的组件都必须通过它提供的 API 来操作资源对象。它以 RESTful 风格的 API 对外提供接口。所有 Kubernetes 资源对象的生命周期维护都是通过调用 API Server 的接口来完成,例如,用户通过 kubectl 创建一个 Pod,即是通过调用 API Server 的接口创建一个 Pod 对象，并存储在 ETCD 集群中。

（2）Controller Manager：集群内部的管理控制中心，主要目的是实现 Kubernetes 集群的故障检测和自动恢复等工作。它包含两个核心组件：Node Controller 和 Replication Controller。其中 Node Controller 负责计算节点的加入和退出，可以通过 Node Controller 实现计算节点的扩容和缩容。Replication Controller 用于 Kubernetes 资源对象 RC 的管理，应用的扩容、缩容以及滚动升级都是由 Replication Controller 来实现。

（3）Kubernetes Scheduler（kube-scheduler）：负责资源的调度（Pod 的调度），相当于公交公司的"调度室"。简而言之，某个 Pod 是在 Node-A 节点上还是 Node-B 节点上都是 kube-scheduler 来进行调度的。

（4）etcd 是高可用键值存储数据库，它是用来存储数据的。在 Kubernetes 集群中，RC、Service、Pod 等一些资源对象都会存储在 etcd 中。

2. node 节点

node 是 kubernetes 集群中除了 master 之外的其他服务器，早期版本称为 Minion。node 可以是物理机也可以是虚拟机，这一点和 master 是一致的。每个 node 上会分配一些工作负载，即 docker 容器。当 node 宕机后，该 node 节点上的应用（Pod）会被转移到其他的 node 节点上。在 node 上也有一些关键的进程，分别是 kubelet、kube-proxy 和 docker engine，具体功能如下：

（1）kubelet：负责本 Node 节点上的 Pod 的创建、修改、监控、删除等 Pod 的全生命周期管理，kubelet 实时向 API Server 发送所在计算 Node 节点的信息。

（2）kube-proxy：实现 Service 的抽象，为一组 Pod 抽象的服务（Service）提供统一接口并提供负载均衡功能。

（3）docker engine：负责节点上容器的创建和管理。

9.2.2 Kubernetes 集群管理

使用 kubectl 常用命令查看信息的方法如下：

（1）Kubernetes 命令语法和用法解析，语法结构如下所示：

```
kubectl [command] [TYPE] [NAME] [flags]
```

command：子命令，用于操作 kubernetes 集群资源对象的命令，如 create、delete、describe、get、apply、logs 等。

TYPE：资源对象的类型，如 Pod、Service、RC、Deployment、Node 等，可以是单数、复数以及简写（pod、pods、po/service、services、svc）。

NAME：资源对象的名称，不指定则返回所有，如"get pod"会返回所有 Pod，"get pod nginx"只返回"Nginx"这个 Pod。

flags：kubectl 子命令的可选参数，如 -n 指定 namespace，-s 指定 apiserver 的 URL。

（2）使用 kubectl api-resources 或 kubectl explain 命令可以获取资源对象类型。资源对象列表如表 9-1 所示。

表 9-1 资源对象列表

名称	简写	名称	简写
componentstatuses	cs	persistentvolumes	pv
daemonsets	ds	persistentvolumeclaims	pvc
deployments	deploy	resourcequotas	quota
events	ev	relocationcontroller	rc
endpoints	ep		secrets
horizontalpodautoscalers	hpa	serviceaccounts	sa
ingresses	ing	services	svc
limitranges	limits	jobs	jobs
nodes	no		
namespace	ns		
pods	po		

使用 kubectl get pods 命令可以获取到所有的 Pod 信息，其实也可以指定某个 Pod 的 Name 进行查看，命令如下：

```
[root@master ~]# kubectl get pods nginx-594c495fdf-5cl9w
```

kubectl 子命令主要包括对资源的创建、删除、查看、修改、配置、运行等功能。

kubectl –help：可以获取所有子命令信息。

kubectl options：可以查看支持的参数，如 --namespace 可以指定所在的 namespace。

（3）kubectl 命令可以用多种格式对结果进行显示，输出格式通过 -o 参数指定，-o 支持的格式如图 9-5 所示：

custom-columns=<spec>	根据自定义列名进行输出，逗号分隔
custom-columns-file=<filename>	从文件中获取自定义列名进行输出
json	以 JSON 格式显示输出结果
jsonpath=<template>	输出 jsonpath 表达式定义的字段信息
jsonpath-file<filename>	输出 jsonpath 表达式定义的字段信息，来源于文件
name	仅输出资源对象的名称
wide	输出更多信息，如输出 Node 名
yaml	以 YAML 格式输出

图 9-5 kubectl 命令格式

```
#kubectl get pod -o wide
#kubectl get pod -o yaml
#kubectl get pod -o
 custom-columns=NAME:.metadata.name,RESC:.metadata.resourceVersion
#kubectl get pod --sort-by=.metadata.name  // 按 Name 排序
```

（4）Kubectl 命令。

创建资源对象，根据 yaml 文件创建 service 和 deployment，命令如下所示：

```
#kubectl create -f nginx-service.yaml -f nginx-deployment.yaml
```

也可以指定一个目录，这样可以一次性根据该目录下所有的 YAML 或 JSON 文件定义资源。

```
#kubectl create -f <directory>
```

查看资源对象，用 kubectl 查看所有 Pod，命令如下所示：

```
#kubectl get pods
```

查看所有的 deployment 和 service，命令如下所示：

```
[root@master ~]# kubectl get deploy,svc
NAME                                      READY     UP-TO-DATE   AVAILABLE   AGE
deployment.extensions/httpd               3/3       3            3           26m
deployment.extensions/nginx-deployment    4/4       4            4           3h51m
NAME                    TYPE        CLUSTER-IP      EXTERNAL-IP   PORT(S)        AGE
service/kubernetes      ClusterIP   10.96.0.1       <none>        443/TCP        4h25m
service/nginx-service   NodePort    10.100.205.95   <none>        80:32600/TCP   89m
```

描述资源对象，显示 Node 的详细信息，命令如下所示：

```
#kubectl describe nodes <node-name>
```

显示 Pod 的详细信息，命令如下所示：

```
#kubectl describe pods <pod-name>
```

显示 deployment 管理的 Pod 信息，命令如下所示：

```
#kubectl describe pods <deployment-name>
```

删除资源对象，基于 YAML 文件删除，命令如下所示：

```
#kubectl delete -f mysql.yaml
```

删除所有包含某个 label 的 pod 和 service，命令如下所示：

```
#kubectl delete pod,svc -l name=<label-name>
```

删除所有 Pod，命令如下所示：

```
#kubectl delete pod --all
```

（5）执行容器的命令。

在 Pod 中执行某个命令，如 date，命令如下所示：

```
#kubectl exec <pod-name> date // pod-name 如果不加，默认会选择第一个 Pod
```

指定 Pod 的某个容器执行命令，命令如下所示：

```
#kubectl exec <pod-name> -c <container-name> date
```

进入到容器里，命令如下所示：

```
#kubectl exec -it <pod-name> -c <container-name> bash
```

（6）查看容器日志。

```
#kubectl logs <pod-name>
```

动态查看容器日志信息，类似于 tail -f 的方式，命令如下所示：

```
#kubectl logs -f <pod-name> -c <container-name>
```

9.2.3 任务：Kubernetes 集群部署

2013 年左右，当时云计算领域的霸主以 AWS 和 OpenStack 为代表，主要提供虚拟机租赁服务，用户购买后用脚本或手工方式在机器上部署应用，部署过程中云端虚拟机与本地环境不一致带来了很多问题。受尽部署痛苦的程序员们打造了以 Cloud Foundry 为代表的开源 PaaS 项目，Cloud Foundry 采用的解决方案是调用操作系统的 Cgroups 和 Namespace 机制为每个应用单独创建隔离环境，即"沙盒"，在"沙盒"中启动应用，这是 PaaS 项目最核心的能力。Cloud

Foundry 打包本地应用的过程令人头疼，Docker 镜像完美解决了 Cloud Foundry 的打包问题，而 Kubernetes 是 Docker 的主流编排工具。

任务执行清单

在本任务中，您将通过实践完成 Kubernetes 双节点集群的部署。

目标

- 掌握 Kubernetes 服务安装。
- 掌握 Kubernetes 集群部署。

重要信息

- 本任务采用 CentOS 7.5_1804 操作系统。
- 虚拟机密码为 000000。
- 提前准备好 K8S 压缩包。
- 关闭 Firewalld 防火墙和 Selinux 服务。

解决方案

1. 基础环境配置

（1）在 Vmware 中安装两台虚拟机，分别作为 master 和 node 节点，下面先登录两个节点更改主机名和主机映射关系。

```
[root@localhost ~]# hostnamectl set-hostname master
[root@localhost ~]# logout
[root@master ~]#
[root@localhost ~]# hostnamectl set-hostname node
[root@localhost ~]# logout
[root@node ~]#
[root@master ~]# vi /etc/hosts
127.0.0.1    localhost localhost.localdomain localhost4 localhost4.localdomain4
::1          localhost localhost.localdomain localhost6 localhost6.localdomain6
192.168.100.10 master
192.168.100.20 node
[root@node ~]# vi /etc/hosts
127.0.0.1    localhost localhost.localdomain localhost4 localhost4.localdomain4
::1          localhost localhost.localdomain localhost6 localhost6.localdomain6
192.168.100.10 master
192.168.100.20 node
```

（2）通过 secureCRT 软件将提供的资源包 K8S.tar.gz 上传到 master 和 node 节点的 /root 目录下，在两个集群节点中创建镜像文件夹，挂载镜像并配置 yum 源，此处以 master 节点为例。

```
[root@master ~]# mkdir /opt/{centos7,k8s}
[root@master ~]# tar -zxvf K8S.tar.gz -C /opt/k8s/
[root@master ~]# mv /etc/yum.repos.d/* /home/
[root@master ~]# mount /dev/cdrom /opt/centos7/
mount: /dev/sr0 is write-protected, mounting read-only
[root@master ~]# vi /etc/yum.repos.d/local.repo
[centos7]
name=centos
baseurl=file:///opt/centos7
enabled=1
```

```
gpgcheck=0
[k8s]
name=k8s
baseurl=file:///opt/k8s/Kubernetes
enabled=1
gpgcheck=0
[root@master ~]# yum repolist
Loaded plugins: fastestmirror
Loading mirror speeds from cached hostfile
repo id                      repo name                              status
centos7                      centos                                 3,971
k8s                          k8s                                    463
```

（3）在两个节点中，关闭 Iptables 服务，并升级操作系统，升级 docker 和 k8s 所需要的内核版本。完成之后重新启动系统，此处以 master 节点为例，命令和结果如下所示：

```
[root@master ~]# iptables -F
[root@master ~]# iptables -Z
[root@master ~]# iptables -X
[root@master ~]# iptables-save
# Generated by iptables-save v1.4.21 on Tue Apr 27 09:18:43 2021
*filter
:INPUT ACCEPT [28:1848]
:FORWARD ACCEPT [0:0]
:OUTPUT ACCEPT [16:1504]
COMMIT
# Completed on Tue Apr 27 09:18:43 2021
[root@master ~]# yum upgrade -y
[root@master ~]# reboot
```

（4）关闭 master 和 node 节点的交换分区，安装时间同步服务器，在 master 节点中的设置如下所示：

```
[root@master ~]# swapoff -a
[root@master ~]# yum install -y chrony
[root@master ~]# vi /etc/chrony.conf
#server 0.centos.pool.ntp.org iburst
#server 1.centos.pool.ntp.org iburst
#server 2.centos.pool.ntp.org iburst
#server 3.centos.pool.ntp.org iburst
local stratum 10
server master iburst
allow all
[root@master ~]# systemctl restart chronyd
[root@master ~]# timedatectl set-ntp true
[root@master ~]# chronyc sources
210 Number of sources = 1
MS Name/IP address         Stratum Poll Reach LastRx Last sample
===============================================================================
^* master                       10   6   377    12   -186ns[-6419ns] +/-    19us
```

下面为 node 节点安装和配置时间同步服务器的详细信息，命令和结果如下所示：

```
[root@node ~]# yum install -y chrony
[root@node ~]# vi /etc/chrony.conf
#server 0.centos.pool.ntp.org iburst
```

```
#server 1.centos.pool.ntp.org iburst
#server 2.centos.pool.ntp.org iburst
#server 3.centos.pool.ntp.org iburst
server 192.168.100.10 iburst
[root@node ~]# systemctl restart chronyd
[root@node ~]# chronyc sources
[root@node ~]# chronyc sources
210 Number of sources = 1
MS Name/IP address         Stratum Poll Reach LastRx Last sample
===============================================================================
^* master                     11    6    17      1   -13us[ -170us] +/-  124ms
```

（5）网络方面，有些应用需要占用端口，而其中一部分应用甚至需要对外提供访问。出于安全方面考虑，代理转发方式相对于直接开放防火墙端口方式更为合适。在 master 和 node 节点中设置内核数据转发和防火墙转发功能，此处以 master 节点为例。

```
[root@master ~]# vi /etc/sysctl.d/K8S.conf
net.ipv4.ip_forward = 1
net.bridge.bridge-nf-call-ip6tables = 1
net.bridge.bridge-nf-call-iptables = 1
[root@master ~]# modprobe br_netfilter
[root@master ~]# sysctl -p /etc/sysctl.d/K8S.conf
net.ipv4.ip_forward = 1
net.bridge.bridge-nf-call-ip6tables = 1
net.bridge.bridge-nf-call-iptables = 1
```

（6）ipvs (IP Virtual Server) 实现了传输层负载均衡，也就是常说的 4 层 LAN 交换，作为 Linux 内核的一部分。ipvs 运行在主机上，在真实服务器集群前充当负载均衡器。

在 master 和 node 节点设置启用 ipvsadm 服务，加载内核的 ipvs 模块。此处以 master 节点为例。

```
[root@master ~]# yum install ipset ipvsadm -y
[root@master ~]# vi /etc/sysconfig/modules/ipvs.modules
#!/bin/bash
modprobe -- ip_vs
modprobe -- ip_vs_rr
modprobe -- ip_vs_wrr
modprobe -- ip_vs_sh
modprobe -- nf_conntrack_ipv4
[root@master ~]# chmod 755 /etc/sysconfig/modules/ipvs.modules
[root@master ~]# bash /etc/sysconfig/modules/ipvs.modules
[root@master ~]# lsmod | grep -e ip_vs -e nf_conntrack_ipv4
nf_conntrack_ipv4       15053  0
nf_defrag_ipv4          12729  1 nf_conntrack_ipv4
ip_vs_sh                12688  0
ip_vs_wrr               12697  0
ip_vs_rr                12600  0
ip_vs                  141432  6 ip_vs_rr,ip_vs_sh,ip_vs_wrr
nf_conntrack           133053  2 ip_vs,nf_conntrack_ipv4
libcrc32c               12644  3 xfs,ip_vs,nf_conntrack
```

2. 集群部署

（1）yum-utils 提供了 yum-config-manager 的依赖包，device-mapper-persistent-data 和 lvm2 需要 devicemapper 存储驱动。

随着 Docker 的不断流行与发展，Docker 组织也开启了商业化之路，Docker 从 17.03 版本之后分为 CE（CommunityEdition）和 EE（EnterpriseEdition）两个版本。Docker CE 是免费的 Docker 产品的新名称，Docker CE 包含了完整的 Docker 平台，非常适合开发人员和运维团队构建容器 App。

在 master 和 node 节点进行服务的安装，此处安装 Docker CE 的版本为 18-09.6，命令如下所示，此处以 master 节点为例，node 节点步骤与其相同。

```
[root@master ~]# yum install -y yum-utils device-mapper-persistent-data lvm2
[root@master ~]# yum install docker-ce-18.09.6 docker-ce-cli-18.09.6 containerd.io -y
```

安装完成，配置 cgroup 驱动，命令和结果如下所示：

```
[root@master ~]# mkdir -p /etc/docker
[root@master ~]# vi /etc/docker/daemon.json
{
  "exec-opts": ["native.cgroupdriver=systemd"]
}
[root@master ~]# systemctl restart docker
[root@master ~]# systemctl daemon-reload
[root@master ~]# docker info | grep Cgroup
 Cgroup Driver: systemd
```

（2）在虚拟机中导入 Kubernetes 服务所需要的相关镜像，使用下列脚本一键导入即可，此处以 master 节点为例，node 节点步骤与其相同。

```
[root@master ~]# cd /opt/k8s/
[root@master k8s]# bash kubernetes_base.sh
```

（3）由于配置双节点集群模式，所以在 master 和 node 节点都需要安装 kubeadm、kubelet、kubectl 包，并启动 kubelet 服务，此处以 master 节点为例，node 节点步骤与其相同，命令和结果如下所示：

```
[root@master ~]# yum install -y kubelet-1.14.1 kubeadm-1.14.1 kubectl-1.14.1
[root@master ~]# systemctl start kubelet
[root@master ~]# systemctl enable kubelet
Created symlink from /etc/systemd/system/multi-user.target.wants/kubelet.service to /usr/lib/systemd/system/kubelet.service.
```

（4）使用 kubeadm init 命令初始化 k8s 服务的 master 节点，并导入环境变量，以便使用 Kubectl 命令用于管理系统及服务。命令如下所示：

```
[root@master ~]# kubeadm init --apiserver-advertise-address 192.168.100.10 --kubernetes-version="v1.14.1"              --pod-network-cidr=10.16.0.0/16 --image-repository=registry.aliyuncs.com/google_containers
```

初始化命令执行后注意结果中加深部分得到的值，每次结果不一样，将其记录下来，类似如下结果：

```
……
kubeadm join 192.168.100.10:6443 --token gtimrr.zq6683wnu5uj3gwb \
--discovery-token-ca-cert-hash sha256:aac6a8658ffd45b895d24917bd6a98d0635f8177b09014a7fb92ee68a0e7cba4
[root@master ~]# mkdir -p $HOME/.kube
[root@master ~]# cp -i /etc/kubernetes/admin.conf $HOME/.kube/config
[root@master ~]# chown $(id -u):$(id -g) $HOME/.kube/config
```

（5）进入压缩包配置文件中，根据提供好的 yaml 文件，加载 k8s 服务的网络插件，创建完成之后使用 kubectl get nodes 查看节点信息，命令和结果如下所示：

```
[root@master ~]# cd /opt/k8s/
[root@master k8s]# bash yaml/kube-flannel.yaml
[root@master ~]# kubectl get nodes
NAME     STATUS   ROLES    AGE     VERSION
master   Ready    master   9m37s   v1.14.1
```

（6）登录 node 节点，从 master 服务器上将 kubeadm init 命令执行之后，提供的 kubeadm join 带有 token 和秘钥的结果，复制到 node 节点下执行，命令和结果如下所示：

```
[root@node ~]# kubeadm join 192.168.100.10:6443 -token
gtimrr.zq6683wnu5uj3gwb     --discovery-token-ca-cert-hash sha256:aac6a8
658ffd45b895d24917bd6a98d0635f8177b09014a7fb92ee68a0e7cba4
```

（7）node 节点集群加入完成之后，使用命令查看进群状态，并查看 Pod 系统服务的状态信息，命令如下，结果如图 9-6 所示。

```
[root@master ~]# kubectl get nodes
NAME     STATUS   ROLES    AGE     VERSION
master   Ready    master   9m37s   v1.14.1
node     Ready    <none>   4m42s   v1.14.1
```

图 9-6　Kubernetes 服务状态

3. Kubernetes 管理平台

（1）在 master 节点，将压缩包中提供的 kubernetes-dashboard.yaml 和 dashboard-adminuser.yaml 文件，使用 kubectl apply 命令进行 Dashboard 安装。命令如下：

```
[root@master ~]# cd /opt/k8s/yaml/
[root@master yaml]# kubectl apply -f kubernetes-dashboard.yaml
[root@master yaml]# kubectl apply -f dashboard-adminuser.yaml
```

如图 9-7 所示，可以看到 dashboard 服务已经正常启动。

图 9-7　Dashboard 服务状态

（2）通过命令检查发现 kubernetes-dashboard 被调度到 Node 节点运行，在 Firefox 浏览器中输入 node 节点地址（Master 也可以访问）https:/192.168.100.20:30000，即可访问 Kubernetes Dashboard，如图 9-8 所示。

（3）单击"高级"→"接受风险并继续"按钮，即可进入 Kubernetes Dashboard 认证界面，如图 9-9 所示。

第 9 章　Kubernetes 服务架构与平台

图 9-8　Kubernetes Dashboard 首次登录　　　图 9-9　Kubernetes Dashboard 认证界面

（4）登录 Kubernetes Dashboard 需要输入令牌，在 master 节点通过以下命令获取访问 Dashboard 的认证令牌，认证后如图 9-10 所示。

图 9-10　Kubernetes 管理平台

```
[root@master ~]# kubectl -n kube-system describe secret $(kubectl -n kube-system
 get secret | grep kubernetes-dashboard-admin-token | awk '{print $1}')
…
token:       eyJhbGciOiJSUzI1NiIsImtpZCI6IiJ9.eyJpc3MiOiJrdWJlcm5ldGVzL3Nlc
nZpY2VhY2NvdW50Iiwia3ViZXJuZXRlcy5pby9zZXJ2aWNlYWNjb3VudC9uYW1lc3BhY2UiOiJrdW
JlLXN5c3RlbSIsImt1YmVybmV0ZXMuaW8vc2VydmljZWFjY291bnQvc2VjcmV0Lm5hbWUiOiJrdWJ
lcm5ldGVzLWRhc2hib2FyZC1hZG1pbi10b2tlbi05cXA5aiIsImt1YmVybmV0ZXMuaW8vc……
```

4. Kuboard 控制平台

（1）Kuboard 是一款免费的 Kubernetes 图形化管理工具，其力图帮助用户快速在 Kubernetes 上落地微服务。登录 master 节点，使用 kuboard.yaml 文件部署 Kuboard。

```
[root@master ~]# kubectl create -f /opt/k8s/yaml/kuboard.yaml
```

（2）在浏览器输入地址 http://192.168.100.20:31000，即可进入 Kuboard 的认证界面，如图 9-11 所示。

图 9-11 Kuboard 认证界面

在图 9-11 中的 Token 文本框内输入令牌后可进入 Kuboard 控制台，如图 9-12 所示。登录到 Kuboard 控制台中可以查看到集群概览，至此 Kubernetes 管理 UI 部署完毕。

图 9-12 Kuboard 控制台

本任务到此结束。

9.3 Kubernetes 编排服务

学习目标

学完本节后，您应能够：
- 了解 Kubernetes 编排文件的语法。
- 了解 Kubernetes 资源与编排文件的编写。
- 掌握 Kubernetes 应用编排与部署。

9.3.1 编排文件的概念和语法

1. 编排文件的概念

Kubernetes master 节点负责维护集群的目标状态。当与 Kubernetes 通信时，使用如 kubectl 的命令行工具，就可以直接与 Kubernetes master 节点进行通信。master 是指管理集群状态的一组进程的集合。通常这些进程都跑在集群中一个单独的节点上，并且这个节点被称为 master 节点。master 节点也可以扩展副本数，来获取更好的可用性及冗余。用于控制 Kubernetes 节点的计算机，所有任务分配都来自于此。

执行请求和分配任务的计算机，由 Kubernetes 主机负责对节点进行控制。集群中的 node 节点（虚拟机、物理机等）都用来运行应用和云工作流的机器。Kubernetes master 节点控制所有的 node 节点，用户很少需要直接和 node 节点通信。

2. 容器编排语法

在 K8S 集群中，容器并非最小的单位，K8S 集群中最小的调度单位是 Pod，容器则被封装在 Pod 之中。由此可知，一个容器或多个容器可以同属于一个 Pod。

以下编写文件 nginx-service.yaml，详细信息如下所示：

```
[root@master ~]# vi nginx-service.yaml
apiVersion: v1
kind: Service
metadata:
  name: nginx-service
  labels:
    app: nginx
spec:
  selector:
    app: nginx
  ports:
  - name: nginx-port
    protocol: TCP
    port: 80
    nodePort: 32600
    targetPort: 80
  type: NodePort
```

使用 kubectl 命令创建 nginx 服务，然后使用命令查看 service 部署结果。

```
[root@master ~]# kubectl apply -f nginx-service.yaml
service/nginx-service created
[root@master ~]# kubectl get services -o wide
NAME            TYPE        CLUSTER-IP       EXTERNAL-IP   PORT(S)        AGE    SELECTOR
kubernetes      ClusterIP   10.96.0.1        <none>        443/TCP        96m    <none>
nginx-service   NodePort    10.110.164.152   <none>        80:32600/TCP   14s    app=nginx
```

（1）YAML 文件编写：YAML 是专门用来写配置文件的语言，非常简洁和强大，使用比 JSON 更方便。它实质上是一种通用的数据串行化格式。在 Kubernetes 中，只需要知道两种结构类型即可：Lists 和 Maps。

YAML 语法规则：

① 大小写敏感。

② 使用缩进表示层级关系。

③ 缩进时不允许使用【Tab】键，只允许使用空格。

④ 缩进的空格数目不重要，只要相同层级的元素左侧对齐即可。

⑤ "#"表示注释,从这个字符一直到行尾,都会被解析器忽略。

使用 YAML 用于 K8S 的定义带来的好处包括:

① 便捷性:不必添加大量参数到命令行中执行命令。

② 可维护性:YAML 文件可以通过源头控制,跟踪每次操作。

③ 灵活性:YAML 文件可以创建比命令行更加复杂的结构。

(2) YAML Maps。顾名思义,Map 指字典,即一个 Key:Value 的键值对信息。例如:

```
---
apiVersion: v1
kind: Service
```

注意:--- 为可选的分隔符,当需要在一个文件中定义多个结构的时候需要使用。上述内容表示有两个键 apiVersion 和 kind,分别对应的值为 v1 和 Pod。

Maps 的 value 既能够对应字符串也能够对应一个 Maps。例如:

```
---
apiVersion: v1
kind: Service
metadata:
  name: nginx-service
  labels:
    app: nginx
```

上述 YAML 文件中,metadata 这个 KEY 对应的值为一个 Map,而嵌套的 labels 这个 KEY 的值又是一个 Map。实际使用中可视情况进行多层嵌套。

YAML 处理器根据行缩进来知道内容之间的关联。上述例子中,使用两个空格作为缩进,但空格的数量并不重要,只是要求至少一个空格并且所有缩进保持一致。例如 name 和 labels 是相同缩进级别,因此 YAML 处理器知道它们属于同一 Map;app 是 lables 的值,因为 app 的缩进量比 labels 更大。

(3) YAML Lists。List 即列表,类似于数组,例如:

```
args
 -beijing
 -shanghai
 -shenzhen
 -anhui
```

可以指定任何数量的项在列表中,每个项的定义以"-"开头,并且与父元素之间存在缩进。在 JSON 格式中,表示如下:

```
{
  "args": ["beijing", "shanghai", "shenzhen", "anhui"]
}
```

当然 Lists 的子项也可以是 Maps,Maps 的子项也可以是 List,例如:

```
---
apiVersion: v1
kind: Pod
metadata:
  name: kube100-site
  labels:
    app: web
```

```
spec:
  containers:
    - name: front-end
      image: nginx
      ports:
        - containerPort: 80
    - name: flaskapp-demo
      image: jcdemo/flaskapp
      ports: 8080
```

如上述文件所示，定义一个 containers 的 List 对象，每个子项都由 name、image、ports 组成，每个 ports 都有一个 KEY 为 containerPort 的 Map 组成。

9.3.2 资源编排与资源文件编写

1. Pod

K8s 有很多技术概念，同时对应很多 API 对象，最重要的也是最基础的是微服务 Pod。Pod 是在 K8s 集群中运行部署应用或服务的最小单元，它是可以支持多容器的。Pod 的设计理念是支持多个容器在一个 Pod 中共享网络地址和文件系统，可以通过进程间通信和文件共享这种简单高效的方式组合完成服务。Pod 本身就是容器，其组成如图 9-13 所示。每一个 Pod 至少有两个容器，Pause 容器和 RC 容器（应用容器），实际上每个 Pod 里面可以有多个应用容器。Pause 容器称为根容器，只有当 Pause 容器死亡才会认为该 Pod 死亡。Pause 容器的 IP 及其挂载的 Volume 资源会共享给该 Pod 下的其他容器。Pause 容器有一个非常重要的功能，它可用来实现网络、数据的共享。

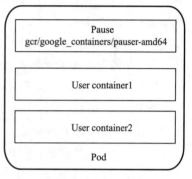

图 9-13 Pod 组成

Pod 是 K8s 集群中所有业务类型的基础，可以看作运行在 K8s 集群中的小机器人，不同类型的业务就需要不同类型的小机器人去执行。目前 K8s 中的业务主要可以分为长期伺服型（long-running）、批处理型（batch）、节点后台支撑型（node-daemon）和有状态应用型（stateful application）；分别对应的小机器人控制器为 Deployment、Job、DaemonSet 和 PetSet。

2. Label

Label 是一个键值对，其中键和值都由用户自定义，Label 可以附加在各种资源对象上，如 Node、Pod、Service、RC 等。一个资源对象可以定义多个 Label，同一个 Label 也可以被添加到任意数量的资源对象上。Label 可以在定义对象时定义,也可以在对象创建完后动态添加或删除。

3. RC

RC（Replication Controller,复制控制器 ）是 K8s 集群中最早的保证 Pod 高可用的 API 对象。通过监控运行中的 Pod 来保证集群中运行指定数目的 Pod 副本。简单地说，它定义了一个期望的场景，即声明某种 Pod 的副本数量在任意时刻都符合某个预期值，RC 定义了如下几个部分：

（1）Pod 期待的副本数。

（2）用于筛选目标 Pod 的 Label Selector。

（3）创建 Pod 副本的模板（template）。

RC 中动态修改 Pod 数量可以用如下命令实现：

```
#kubectl scale rc <rc name> --replicas=n
```

利用动态修改 Pod 的副本数，可以实现应用的动态升级（滚动升级）：

（1）以新版本的镜像定义新的 RC，但 Pod 要和旧版本保持一致（由 Label 决定）。

（2）新版本每增加一个 Pod，旧版本就减少一个 Pod，始终保持固定的值。

（3）最终旧版本 Pod 为 0，全部为新版本。

删除 RC 后，RC 对应的 Pod 也会随之删除。删除 RC 命令如下所示：

```
#kubectl delete rc <rc name>
```

但是 Service 还是存在的，需要手动进行删除。删除 svc 命令如下所示：

```
#kubectl delete svc <svc name>
```

4. Deployment

Deployment（部署）表示用户对 K8s 集群的一次更新操作，和 RC 类似，不过 Deployment 是新版的 RC，它比 RC 更高级一些，Deployment 完全可以替代 RC。Deployment 是在 1.2 版本中引入的概念，目的是解决 Pod 编排问题，在内部使用了 RS（Replicat Set，副本集），新一代 RC，提供同样的高可用能力，区别主要在于 RS 后来居上，能支持更多种类的匹配模式。副本集对象一般不单独使用，而是作为 Deployment 的理想状态参数使用。以 K8s 的发展方向，未来对所有长期伺服型业务的管理，都会通过 Deployment 来管理。

5. HPA（Horizontail Pod Autoscaler）

在 Kubernetes1.1 版本时，Kubernetes 官方发布了 HPA，它可以实现 Pod 的动态扩容、缩容，属于一种 Kubernetes 资源对象。它通过追踪分析 RC 控制的所有目标 Pod 的负载变化情况，来决定是否需要针对性地调整目标 Pod 的副本数，这是 HPA 实现的原理。

用命令行的方式定义 HPA，例如，HPA 控制的目标对象是一个名叫 php-apache 的 Deployment 里的 Pod 副本，当 CPU 平均值超过 90% 时就会扩容，Pod 副本数控制范围是 1～10，具体命令如下：

```
# kubectl autoscale deployment php-apache --cpu-percent=8080 --min=1
--max=10
```

6. Service

Service 是 kubernetes 中最核心的资源对象之一，没有 Service 自然而然也就没有了应用，就不能访问服务了。Service 可以理解为微服务架构中的一个微服务。RC、RS 和 Deployment 只是保证了支撑服务的微服务 Pod 的数量，但是没有解决如何访问这些服务的问题。

简而言之，一个 Service 本质上是一组 Pod 组成的一个集群，由于 Service 和 Pod 之间是通过 Label 串起来的，相同的 Service 和 Pod 的 Label 是一样的。同一个 Servic 下的所有 Pod 是通过 kube-proxy 实现负载均衡的，而每个 Service 都会分配一个全局唯一的虚拟 IP，也就是通过 kubectl get svc 命令看到的 CLUSTER-IP。在该 Service 整个生命周期内，CLUSTER-IP 是不会改变的，而在 Kubernetes 中还有一个 DNS 服务，它把 Service 的 Name 和 CLUSTER-IP 映射起来。

通过 kubectl get endpoints 命令可以查看 Pod 的 IP 地址及端口号，Pod 的 endpoints 下方会显示多个 IP 地址，是由于对应 Pod 存在多个副本数造成的。

通过 kubectl get svc <service-name> -o yaml 命令可以查看 Service 分配的 Cluster-IP，对于 Cluster-IP 有如下限制：

（1）Cluster-IP 无法被 Ping 通，因为没有实体网络来响应。

（2）Cluster-IP 和 Service Port 组成了一个具体的通信端口，单独的 Cluster-IP 不具备 TCP/IP 通信基础，它们属于一个封闭的空间。

（3）在 Kubernetes 集群中，Node IP、Pod IP、Cluster IP 之间的通信，采用的是 Kubernetes 自己设计的一套编程方式的特殊路由规则。

要想直接和 Service 进行通信，需要一个 NodePort。NodePort 是在 Service 的 YAML 文件中定义，它实质上是把 Cluster-IP 的 Port 映射到了 Node IP 的 NodePort 上了。而 NodePort 端口号需要大于 30000 才可以。

7. Volume

Volume（存储卷）是 Pod 中能够被多个容器访问的共享目录，Kubernetes 当中的 Volume 和 Docker 中的 Volume 不一样，主要有如下几个方面：

（1）Kubernetes 的 Volume 定义在 Pod 上，然后被一个 Pod 里面的多个容器挂载到具体的目录下。

（2）Kubernetes 的 Volume 与 Pod 生命周期相同，但与容器的生命周期没有关系。当容器终止后启动时，Volume 中的数据并不会丢失。

（3）Kubernetes 支持多种类型的 Volume，如 GlusterFS、CEPH 等先进的分布式文件系统。

定义并使用 Volume 只需要在定义 Pod 的 YAML 配置文件中指定 Volume 相关配置即可。

8. PV 和 PVC

PV（Persistent Volume，持久存储卷）可以理解成 Kubernetes 集群中某个网络存储中对应的一个网络存储，它与 Volume 类似，但是有如下区别：

（1）PV 只能是网络存储，不属于任何 Node，但是可以在每个 Node 上访问到。

（2）PV 并不是定义在 Pod 上，而是独立于 Pod 之外。

（3）PV 目前只存在的几种类型：GCE Persistent Disk、NFS、RBD、iSCSCI、AWS、ElasticBlockStore、GlusterFS。

NFS 类型 PV 定义如下所示：

```
apiVersion: v1
kind: PersistentVolume
metadata:
  name: mysql-persistent-storage
  namespace: default
  labels:
    app: wordpress
spec:
  accessModes:
  - ReadWriteOnce
  capacity:
    storage: 20Gi
  nfs:
    path: /data-mysql
    server: master
  persistentVolumeReclaimPolicy: Recycle
  storageClassName: nfs
```

其中 accessModes 是一个重要的属性，目前有以下类型：

（1）ReadWriteOnce：读写权限，并且只能被单个 node 挂载。

（2）ReadOnlyMany：只读权限，允许被多个 node 挂载。

（3）ReadWriteMany：读写权限，允许被多个 node 挂载。

注意：如果某个 Pod 想申请某种条件的 Persistent Volume，首先需要定义一个 PVC（Persistent Volume Claim，持久存储卷声明）对象。也就是说如果想用 PV 就需要先有 PVC。定义 PVC 示例如下所示：

```yaml
apiVersion: v1
kind: PersistentVolumeClaim
metadata:
  name: mysql-pv-claim
  namespace: default
  labels:
    app: wordpress
spec:
  accessModes:
  - ReadWriteOnce
  resources:
    requests:
      storage: 20Gi
  storageClassName: nfs
  volumeName: mysql-persistent-storage
```

PVC 可以自动和 PV 关联起来。有了 PVC 后，在 Pod 的 Volume 中去引用 PVC。先定义 PV，PV 会和 PVC 自动关联起来，最后在 Pod 或者 RC、Deployment 里面去引用 PVC 就可以了。Pod 的 Volume 定义引用 PVC 示例如下所示：

```yaml
volumes:
      -name: mysql-pv
        persistentVolumeClaim:
          claimName: mysql-pv-claim
```

9. NameSpace

NameSpace（名字空间）为 K8s 集群提供虚拟的隔离作用，K8s 集群初始有两个名字空间，分别是默认名字空间 default 和系统名字空间 kube-system，除此以外，管理员可以创建新的名字空间满足需要。使用 kubectl get ns 可以查看所有的 namespace。ns 是简写，全拼是 kubectl get namespace，命令如下所示：

```
# kubectl get ns
```

NameSpace 定义如下所示：

```yaml
apiVersion: v1
kind: Namespace
metadata:
  name: test_ns
```

创建好 NameSpace 后，使用 kubectl create -f <namespace-files-name>.yaml 命令即可创建 NameSpace 命名空间。创建好 NameSpace 后，再定义 Pod 指定 namespace 即可。示例如下所示：

```yaml
apiVersion: v1
kind: Pod
metadata:
  name: busybox
  namespace: humingzhe
```

```
    spec:
    containers:
    - image: busybox
      command:
      - sleep
      - "500"
      name: busybox
```

9.3.3 任务：Kubernetes 编排案例

任务执行清单

在本任务中，您将通过 Kubernetes 服务编排 Nginx 服务案例。

目标

- 掌握 Deployment 编写。
- 掌握 Nginx 服务编排。
- 掌握 Pod 副本数的扩容。

Kubernetes 是 Google 开源的一个容器编排引擎，它支持自动化部署、大规模可伸缩、应用容器化管理。在生产环境中部署一个应用程序时，通常要部署该应用的多个实例以便对应用请求进行负载均衡。

重要信息

- 本任务采用 CentOS 7.5_1804 操作系统。
- 虚拟机密码为 000000。
- 所有节点已安装好 docker-ce。
- 所有节点已部署 Kubernetes。
- 关闭 Firewalld 防火墙和 Selinux 服务。

解决方案

1. Deployment 编写

（1）登录 master 节点，在 root 目录下编写 yaml 编排文件，编写文件 nginx-deployment.yaml，文件的详细信息如下所示：

```
[root@master ~]# vi nginx-deployment.yaml
apiVersion: apps/v1
kind: Deployment
metadata:
  name: nginx-deployment
  labels:
    app: nginx
spec:
  replicas: 1
  selector:
    matchLabels:
      app: nginx
  template:
    metadata:
      labels:
```

```
        app: nginx
  spec:
    containers:
    - name: nginx
      image: nginx
      imagePullPolicy: IfNotPresent
```

其中，字段说明如下所示。

metadata 指元数据，即 Deployment 的一些基本属性和信息。

name: nginx-deployment：指 Deployment 的名称。

labels 指标签，可以灵活定位一个或多个资源，其中 key 和 value 均可自定义，可以定义多组。

app nginx：为该 Deployment 设置 key 为 app、value 为 nginx 的标签。

（2）使用 kubectl apply 命令，通过 -f 参数指定文件信息，创建服务，命令和结果如下所示：

```
[root@master ~]# kubectl apply -f nginx-deployment.yaml
deployment.apps/nginx-deployment created
```

（3）查看 Deployment 服务的结果，命令和结果如下所示：

```
[root@master ~]# kubectl get deployments
NAME               READY   UP-TO-DATE   AVAILABLE   AGE
nginx-deployment   1/1     1            1           14s
```

使用命令继续查看 Pod 的状态信息，命令和结果如下所示：

```
[root@master ~]# kubectl get pods
NAME                                READY   STATUS    RESTARTS   AGE
nginx-deployment-7cffb9df96-p5m95   1/1     Running   0          21m
```

2. Servcie 文件编写

（1）登录 master 节点，在 root 目录下编写 service 编排文件，编写文件 nginx-service.yaml，文件的详细信息如下所示：

```
[root@master ~]# vi nginx-service.yaml
apiVersion: v1
kind: Service
metadata:
  name: nginx-service
  labels:
    app: nginx
spec:
  selector:
    app: nginx
  ports:
  - name: nginx-port
    protocol: TCP
    port: 80
    nodePort: 32600
    targetPort: 80
  type: NodePort
```

（2）使用 kubectl 命令创建 nginx 服务，然后使用命令查看 service 部署结果，命令和结果如下所示：

```
[root@master ~]# kubectl apply -f nginx-service.yaml
service/nginx-service created
```

```
[root@master ~]# kubectl get services -o wide
NAME            TYPE        CLUSTER-IP      EXTERNAL-IP   PORT(S)         AGE   SELECTOR
kubernetes      ClusterIP   10.96.0.1       <none>        443/TCP         96m   <none>
nginx-service   NodePort    10.110.164.152  <none>        80:32600/TCP    14s   app=nginx
```

（3）在浏览器中输入 http://192.168.100.10:32600，即可访问我们一键部署的 Nginx 服务了，结果如图 9-14 所示。

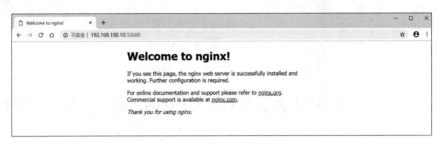

图 9-14 Nginx 部署成功

3. 服务扩容

（1）伸缩是指在线增加或者减少 Pod 的副本数，部署 nginx-deployment 初始的 1 个副本，如下所示：

```
[root@master ~]# kubectl get pods
NAME                                    READY   STATUS    RESTARTS   AGE
nginx-deployment-7cffb9df96-bghdf       1/1     Running   0          160m
```

（2）修改 nginx-deployment.yaml 部署文件，将副本 replicas：1 中的调整成 3，如下所示：

```
[root@master ~]# vi nginx-deployment.yaml
apiVersion: apps/v1
kind: Deployment
metadata:
  name: nginx-deployment
  labels:
    app: nginx
spec:
  replicas: 3
  selector:
    matchLabels:
      app: nginx
  template:
    metadata:
      labels:
        app: nginx
    spec:
      containers:
      - name: nginx
        image: nginx
        imagePullPolicy: IfNotPresent
```

重新部署，再次执行 kubectl apply 命令，命令如下所示：

```
[root@master ~]# kubectl apply -f nginx-deployment.yaml
deployment.apps/nginx-deployment configured
```

再次查看伸缩后的结果，如下所示：

```
[root@master ~]# kubectl get pods -o wide
NAME                                READY   STATUS    RESTARTS   AGE
IP           NODE    NOMINATED NODE   READINESS GATES
nginx-deployment-7cffb9df96-4tg6p   1/1     Running   0          6s
10.16.1.5    node    <none>           <none>
nginx-deployment-7cffb9df96-p5m95   1/1     Running   0          8m45s
10.16.1.4    node    <none>           <none>
nginx-deployment-7cffb9df96-sqfwj   1/1     Running   0          6s
10.16.1.6    node    <none>           <none>
```

本任务到此结束。

9.4 开放研究任务：Kubernetes 编排 WordPress

某公司将基于 Kubernetes 服务平台，使用 Deployment、NFS、PV、PVC 等方法开发一套数据持久化存储的资源平台，研发部决定使用微服务架构。

任务执行清单

在本任务中，您将基于 Kubernetes 服务编排部署 WordPress+MySQL 的案例，使用 NFS 挂载完成共享存储，实现数据持久化存储的案例。

（1）基础环境配置。

（2）PV 与 PVC 创建。

（3）平台部署。

目标

- 掌握 Kubernetes 服务编排。
- 掌握 WordPress 服务部署。
- 掌握 NFS 服务配置与挂载。
- 掌握 WordPress 数据持久化存储。

重要信息

- 本任务采用 CentOS 7.5_1804 操作系统。
- 虚拟机密码为 000000。
- 所有节点已安装好 docker-ce。
- 所有节点已部署 Kubernetes。
- 关闭 Firewalld 防火墙和 Selinux 服务。

解决方案

1. 基础环境配置

在扩大系统的规模时，WordPress 应用服务的 Pod 可以增加运行的拷贝数量，遇到故障时 Pod 可以转移到其他节点。为了浮动节点都能够访问统一的存储，我们使用 NFS 来建立网络存储服务，当然在云数据中心也可以使用云计算服务商提供的存储卷。对于规模较大的站点，也可以自己利用或部署 Rook（Ceph）分布式存储系统。

（1）登录到 master 节点，使用 yum 源安装本地 NFS 服务，创建两个文件夹，mysql 目录用

于映射容器里面的 /var/lib/mysql，wordpress 目录用于映射容器里面的 /var/www/html。这里需保证 nfs-utils 安装到所有 master 和 node 中，否则容器挂载 NFS 时会报错。

```
[root@master ~]# yum install -y nfs-utils rpcbind
[root@master ~]# mkdir /data-mysql /data-wordpress
[root@master ~]# chmod 777 /data-mysql
[root@master ~]# chmod 777 /data-wordpress
```

（2）在 NFS 配置文件中，设置需要共享的目录信息，配置完成使用 exportfs 进行服务的测试，命令和结果如下所示：

```
[root@master ~]# vi /etc/exports
/data-mysql *(rw,no_root_squash,sync)
/data-wordpress *(rw,no_root_squash,sync)
[root@master ~]# exportfs -r
[root@master ~]# exportfs -v
/data-mysql     <world>(sync,wdelay,hide,no_subtree_check,sec=sys,rw,no_root_squash,no_all_squash)
/data-wordpress           <world>(sync,wdelay,hide,no_subtree_check,sec=sys,rw,no_root_squash,no_all_squash)
```

（3）启动 NFS 服务，在 node 客户端进行验证，命令和结果如下所示：

```
[root@master ~]# systemctl enable nfs rpcbind
Created symlink from /etc/systemd/system/multi-user.target.wants/nfs-server.service to /usr/lib/systemd/system/nfs-server.service.
[root@master ~]# systemctl start nfs rpcbind
[root@master ~]# showmount -e 192.168.100.10
Export list for 192.168.100.10:
/data-wordpress *
/data-mysql     *
```

2. PV 与 PVC 创建

（1）Persistent volume 配置，登录 master 节点，为 data-mysql 源码存储创建 Persistent volume，编写 mysql 的案例 mysql-pv.yaml，详细信息如下所示：

```
[root@master ~]# vi mysql-pv.yaml
apiVersion: v1
kind: PersistentVolume
metadata:
  name: mysql-persistent-storage
  namespace: default
  labels:
    app: wordpress
spec:
  accessModes:
  - ReadWriteOnce
  capacity:
    storage: 20Gi
  nfs:
    path: /data-mysql
    server: master
  persistentVolumeReclaimPolicy: Recycle
  storageClassName: nfs
```

使用 kubectl apply 命令，指定 mysql-pv.yaml 文件进行 PV 数据卷的创建，最后查看当前 pv 的状态信息，命令和结果如下所示：

```
[root@master ~]# kubectl apply -f mysql-pv.yaml
persistentvolume/mysql-persistent-storage created
[root@master ~]# kubectl get pv
 NAME                               CAPACITY   ACCESS MODES   RECLAIM POLICY
STATUS      CLAIM    STORAGECLASS    REASON    AGE
 mysql-persistent-storage   20Gi         RWO             Recycle
Available              nfs                       13s
```

（2）Persistent volume 配置，登录 master 节点，为 data-wordpress 源码存储创建 Persistent volume，编写 mysql 的案例 mysql-pv.yaml，详细信息如下所示：

```
[root@master ~]# vi wordpress-pv.yaml
apiVersion: v1
kind: PersistentVolume
metadata:
  name: wordpress-persistent-storage
  namespace: default
  labels:
    app: wordpress
spec:
  accessModes:
  - ReadWriteOnce
  capacity:
    storage: 20Gi
  nfs:
    path: /data-wordpress
    server: master
  persistentVolumeReclaimPolicy: Recycle
  storageClassName: nfs
```

使用 kubectl apply 命令，指定 wordpress-pv.yaml 文件进行 PV 数据卷的创建，最后查看 PV 状态信息，命令如下所示，PV 状态信息如图 9-15 所示。

```
[root@master ~]# kubectl apply -f wordpress-pv.yaml
persistentvolume/wordpress-persistent-storage created
```

图 9-15　PV 状态信息

（3）创建存放 data-mysql 的 PVC，创建 PVC 需要在 PV 已创建好的基础上操作，且在查看到 PV 状态信息为 Available 时，编写 mysql-pvc.yaml 文件，详细信息如下所示：

```
[root@master ~]# vi mysql-pvc.yaml
apiVersion: v1
kind: PersistentVolumeClaim
metadata:
  name: mysql-pv-claim
  namespace: default
  labels:
    app: wordpress
spec:
  accessModes:
```

```
    - ReadWriteOnce
  resources:
    requests:
      storage: 20Gi
  storageClassName: nfs
  volumeName: mysql-persistent-storage
```

使用 kubectl apply 命令，指定 mysql-pvc.yaml 文件进行 PVC 数据卷的绑定，最后查看 PVC 状态信息，命令和结果如下所示：

```
[root@master ~]# kubectl apply -f mysql-pvc.yaml
persistentvolumeclaim/mysql-pv-claim created
[root@master ~]# kubectl get pvc
NAME             STATUS   VOLUME                     CAPACITY   ACCESS MODES   STORAGECLASS   AGE
mysql-pv-claim   Bound    mysql-persistent-storage   20Gi       RWO            nfs            7s
```

（4）创建存放 data-wordpress 的 PVC，创建 PVC 需要在 PV 已创建好的基础上操作，且在查看到 PV 状态信息为 Available 时，编写 wordpress-pvc.yaml 文件，详细信息如下所示：

```
[root@master ~]# vi wordpress-pvc.yaml
apiVersion: v1
kind: PersistentVolumeClaim
metadata:
  name: wp-pv-claim
  namespace: default
  labels:
    app: wordpress
spec:
  accessModes:
    - ReadWriteOnce
  resources:
    requests:
      storage: 20Gi
  storageClassName: nfs
  volumeName: wordpress-persistent-storage
```

使用 kubectl apply 命令，指定 mysql-pvc.yaml 文件进行 PVC 数据卷的绑定，最后使用命令查看 PV 和 PVC 的信息，命令如下所示，结果如图 9-16 所示。

```
[root@master ~]# kubectl apply -f wordpress-pvc.yaml
persistentvolumeclaim/wp-pv-claim created
```

```
[root@master ~]# kubectl get pv
NAME                           CAPACITY   ACCESS MODES   RECLAIM POLICY   STATUS   CLAIM                    STORAGECLASS   REASON   AGE
mysql-persistent-storage       20Gi       RWO            Recycle          Bound    default/mysql-pv-claim   nfs                     4h35m
wordpress-persistent-storage   20Gi       RWO            Recycle          Bound    default/wp-pv-claim      nfs                     4h34m
[root@master ~]#
[root@master ~]# kubectl get pvc
NAME             STATUS   VOLUME                         CAPACITY   ACCESS MODES   STORAGECLASS   AGE
mysql-pv-claim   Bound    mysql-persistent-storage       20Gi       RWO            nfs            4h33m
wp-pv-claim      Bound    wordpress-persistent-storage   20Gi       RWO            nfs            4h32m
```

图 9-16　PV 和 PVC 状态信息

3. 平台部署

（1）Deployment 配置，部署 mysql deployment with PVC，编写 mysql-deployment.yaml 文件，详细信息如下所示：

```
[root@master ~]# vi mysql-deployment.yaml
kind: Service
```

```yaml
apiVersion: v1
metadata:
  labels:
    app: wordpress
  name: wordpress-mysql
spec:
  ports:
    - port: 3306
  selector:
    app: wordpress
    tier: mysql
  clusterIP: None
---
kind: Deployment
apiVersion: apps/v1
metadata:
  labels:
    app: wordpress
  name: wordpress-mysql
spec:
  replicas: 1
  revisionHistoryLimit: 10
  selector:
    matchLabels:
      app: wordpress
  template:
    metadata:
      labels:
        app: wordpress
        tier: mysql
    spec:
      containers:
        - name: wordpress
          image: mysql:5.6
          imagePullPolicy: IfNotPresent
          ports:
            - containerPort: 3306
              name: mysql
          env:
            - name: MYSQL_ROOT_PASSWORD
              value: '123456'
          volumeMounts:
            - mountPath: /var/lib/mysql
              name: mysql-pv
      volumes:
        - name: mysql-pv
          persistentVolumeClaim:
            claimName: mysql-pv-claim
      nodeName: master
```

使用 kubectl apply 命令，指定 mysql-deployment.yaml 文件进行服务的创建与配置，命令和结果如下所示：

```
[root@master ~]# kubectl apply -f mysql-deployment.yaml
```

```
service/wordpress-mysql created
deployment.apps/wordpress-mysql created
```

（2）Deployment 配置，部署 wordpress deployment with PVC，编写 wordpress-deployment.yaml 文件，详细信息如下所示：

```
[root@master ~]# vi wordpress-deployment.yaml
kind: Service
apiVersion: v1
metadata:
  labels:
    app: wordpress
  name: wordpress
spec:
kind: Service
apiVersion: v1
metadata:
  labels:
    app: wordpress
  name: wordpress
spec:
  type: NodePort
  ports:
    - port: 80
      nodePort: 31000
  selector:
    app: wordpress
    tier: frontend
---
kind: Deployment
apiVersion: apps/v1
metadata:
  labels:
    app: wordpress
  name: wordpress
spec:
  replicas: 1
  revisionHistoryLimit: 10
  selector:
    matchLabels:
      app: wordpress
  template:
    metadata:
      labels:
        app: wordpress
        tier: frontend
    spec:
      containers:
        - name: wordpress
          image: 192.168.100.10/wordpress:latest
          imagePullPolicy: IfNotPresent
          ports:
            - containerPort: 80
              name: wordpress
```

```
          env:
            - name: WORDPRESS_DB_HOST
              value: wordpress-mysql
            - name: WORDPRESS_DB_PASSWORD
              value: '123456'
          volumeMounts:
            - mountPath: /var/www/html
              name: wordpress-pv
      volumes:
        - name: wordpress-pv
          persistentVolumeClaim:
            claimName: wp-pv-claim
      nodeName: master
```

使用 kubectl apply 命令,指定 wordpress-deployment.yaml 文件进行服务的创建与配置,命令和结果如下所示:

```
[root@master ~]# kubectl apply -f wordpress-deployment.yaml
service/wordpress created
deployment.apps/wordpress created
```

(3) 两个服务都部署完成之后,打开浏览器输入 https://192.168.100.10:31000,可以看见图 9-17 所示的 WordPress 初始化界面,添加相关信息,安装 WordPress 服务,图 9-18 所示为登录到 WordPress 博客主页。

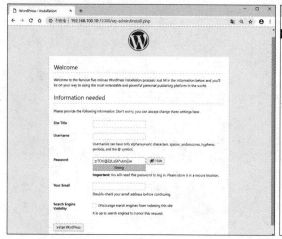

图 9-17　WordPress 初始化界面　　　　图 9-18　WordPress 博客主页

本任务到此结束。

小　　结

在本章中,您已经学会:
- Kubernetes 服务的原理与架构。
- Kubernetes 集群架构的部署。
- Kubernetes 编排案例的部署。
- Kubernetes 编排 WordPress。